D1756963

DETAILS FOR BORROWER'S SLIPS:

Auth/Ed: THOMASON

Title: Ductile fracture of metals

Vol: Copy No: 2

Stamp indicates date for RETURN. Fines for late returns will be charged in accordance with the regulations.
Books required by another reader will be recalled after 3 days.

Ductile Fracture of Metals

Pergamon Titles of Related Interest

ASHBY
Engineering Materials 1 and 2

BROOK & HANSTEAD
Reliability in Non-Destructive Testing

FAN & MURAKAMI
Advances in Constitutive Laws for Engineering Materials, 2-Volume Set

FU
Metallurgy and Materials Science of Tungsten, Titanium, Rare Earths and Antimony, 2-Volume Set

HEARN
Mechanics of Materials, 2-Volume Set, 2nd Edition

KELLY
Concise Encyclopedia of Composite Materials

KETTUNEN ET AL
Strength of Metals and Alloys, 3-Volume Set

KHAN & TOKUDA
Advances in Plasticity

KING
Surface Treatment and Finishing of Aluminium

SALAMA ET AL
Advances in Fracture Research, 6-Volume Set

TAYA & ARSENAULT
Metal Matrix Composites: Thermomechanical Behaviour

Pergamon Related Journals
(*free specimen copy gladly sent on request*)

Acta Metallurgica
Canadian Metallurgical Quarterly
Engineering Fracture Mechanics
Fatigue and Fracture of Engineering Materials and Structures
International Journal of Impact Engineering
International Journal of Solids and Structures
Journal of the Mechanics and Physics of Solids
Materials and Society
Materials Research Bulletin
Physics of Metals and Metallography
Progress in Materials Science
Scripta Metallurgica
Solid State Communications
Welding in the World

Ductile Fracture of Metals

P. F. THOMASON
University of Salford, UK

PERGAMON PRESS
Member of Maxwell Macmillan Pergamon Publishing Corporation

OXFORD · NEW YORK · BEIJING · FRANKFURT
SÃO PAULO · SYDNEY · TOKYO · TORONTO

U.K.	Pergamon Press plc, Headington Hill Hall, Oxford OX3 0BW, England
U.S.A.	Pergamon Press, Inc., Maxwell House, Fairview Park, Elmsford, New York 10523, U.S.A.
PEOPLE'S REPUBLIC OF CHINA	Pergamon Press, Room 4037, Qianmen Hotel, Beijing, People's Republic of China
FEDERAL REPUBLIC OF GERMANY	Pergamon Press GmbH, Hammerweg 6, D-6242 Kronberg, Federal Republic of Germany
BRAZIL	Pergamon Editora Ltda, Rua Eca de Queiros, 346, CEP 04011, São Paulo, Brazil
AUSTRALIA	Pergamon Press (Australia) Pty Ltd, PO Box 544, Potts Point, NSW 2011, Australia
JAPAN	Pergamon Press, 5th Floor, Matsuoka Central Building, 1-7-1 Nishishinjuku, Shinjuku-ku, Tokyo 160, Japan
CANADA	Pergamon Press Canada Ltd, Suite No 271, 253 College Streeet, Toronto, Ontario, Canada M5T 1R5

TO PAULINE

CONTENTS

PREFACE

The various mechanisms of microvoid nucleation, growth and coalescence that lead to the formation of ductile-fracture surfaces in metals can only be understood in terms of models that operate at structural levels varying from the sub-microscopic level, where dislocation models apply, to the microscopic and macroscopic levels, where plastic-continuum models are required. Consequently, a strong interdisciplinary approach is needed in research on the problem of ductile fracture, bringing together results on the mechanics and physics of the fracture processes obtained by metallurgists, physicists, mathematicians and engineers. In writing this book I have aimed at drawing together, in a concise form, a wide range of research results from these associated disciplines to give an account of the progress that has been made in understanding the fundamental mechanisms of ductile fracture. A book of this type seems particularly timely because it is now becoming clear that further progress in developing more reliable design methods in elastic/plastic fracture mechanics will require crack-stability models which take into account the fundamental mechanisms of the ductile-fracture process. The need for a more fundamental approach also applies to the related problem of developing more reliable models for estimating the limits of metalforming ductility in modern structural materials.

This book should be of interest to final-year honours degree undergraduates, postgraduate students, research scientists and engineers from the disciplines of materials science and engineering, metallurgy, aeronautical, mechanical and manufacturing engineering, who require a detailed account of current understanding of the mechanics and physics of ductile fracture in metals. An important feature of the book is that Chapter 3 locates and discusses the conceptual errors contained in the formulation of the Berg-Gurson-Tvergaard model of ductile fracture, and goes on to show how the plastic limit-load model of ductile fracture avoids these conceptual problems by the introduction of a dual ("strong" and "weak") dilational-plastic response. The dual plastic-constitutive response contains an allowance for the actual "presence" of voids in a continuum model of ductile fracture, which is shown to be an essential requirement for any valid model of ductile fracture in metals. In Chapter 8 the book also confronts the conceptual weaknesses in the J-controlled crack growth approach to elastic/plastic fracture mechanics, and it is shown that an alternative incremental-plastic formulation of the crack-stability problem is possible.

In order to keep this book to a reasonable length I have assumed that the reader will have a working knowledge of the basic mathematical theory of plasticity, including slip-line field theory and upper-and lower-bound principles. The relevant plasticity theory is covered by numerous books, but R. Hill's "The Mathematical Theory of Plasticity" and L.M. Kachanov's "Fundamentals of the Theory of Plasticity" would form particularly appropriate reading for those requiring an introduction to the necessary background material.

The planning of this book and the preparation of the first four chapters were carried out when I was on sabbatical leave as a Visiting Professor in the Department of Materials Science and Mineral Engineering, University of California, Berkeley, U.S.A. I would therefore like to take this opportunity of thanking Professor Rob. Ritchie, Professor of Metallurgy at U.C. Berkeley and Director of Research at the Center for Advanced Materials, Lawrence Berkeley Laboratories, for his generous hospitality throughout my twelve-months visit to Berkeley; Professor Ritchie and a number of colleagues at Berkeley kindly read the manuscript of the first four chapters of the book and made valuable comments. I would also like to thank my colleagues and students at Berkeley who helped to make my visit so interesting and enjoyable. In connection with my visit to the U.S.A., I am most grateful to the University of Salford for awarding me a twelve-months period of study leave for the 1987/88 academic year, and the Royal Society for the award of a U.S. Research Grant towards my travel costs.

Finally, I would like to express my gratitude to the many authors and publishers who have so generously given me permission to reproduce results from their publications; the necessary acknowledgements have been made at appropriate points in the text.

October 1989.   P.F.T.

CHAPTER 1

Mechanical and Physical Principles of Plasticity and Ductile Fracture

1.1 Physical Mechanisms of Plastic Flow and the Plastic-Yield Criterion

The plastic deformation of metals is the result of the movement of dislocations along the slip planes of the crystalline lattice, under the action of resolved slip-plane shear stresses [1,2]. Crystallographic dislocations are composed of varying proportions of edge and screw components [1], and in an annealed crystal they exist as low-density networks throughout the crystalline lattice. When a polycrystalline aggregate is subjected to increasing applied stress, plastic deformation is brought about by the rapid multiplication of mobile dislocations to give a crystalline structure of increasing dislocation density [1,2]. The rapid multiplication of dislocations usually leads to an increasingly strong interaction between the individual dislocation stress fields and those of neighbouring dislocations thus impeding the further movement of the dislocations and requiring increasing applied-stress levels to bring about continuing plastic deformation. This process is responsible for the well known work-hardening effect in polycrystalline metal aggregates [1-3]. An important physical feature of the large plastic deformation or 'flow' of a polycrystalline metal is that the vast multiplication of dislocations, which accompanies the process, results in only a very small overall elastic dilation of the crystalline lattice; and this is negligible in comparison to very large plastic strains that occur. This effect is a direct result of the slip or glide nature of dislocation movements where the crystalline lattice is merely being displaced in a direction tangential to the slip plane, and no matter how many dislocations traverse a slip plane there is no change in the volume of the crystalline lattice over and above the very small elastic dilations accompanying each pile-up of dislocations. It follows from this result that, for all practical purposes, the plastic flow of polycrystalline metals can be regarded as essentially incompressible, and this effect can be represented mathematically in terms of the components of the plastic strain -increment tensor $d\epsilon_{ij}^p$ [4,5] by the relation:

1

$$d\epsilon_{ii}^p = d\epsilon_x^p + d\epsilon_y^p + d\epsilon_z^p = d\epsilon_1^p + d\epsilon_2^p + d\epsilon_3^p = 0, \qquad (1.1)$$

where the repeated suffix implies summation and $d\epsilon_1^p$, $d\epsilon_2^p$, $d\epsilon_3^p$ are the principal components of $d\epsilon_{ij}^p$. The incompressibility equation (1.1) is generally valid at all stages of a plastic flow process in metals and has important consequences in both the mathematical theory of plasticity and the theory of ductile fracture; these aspects of the metal plasticity and fracture problem will be discussed in detail at appropriate stages of the following work.

The general plastic strain-increment tensor $d\epsilon_{ij}^p$ has nine components but, from the symmetry of the tensor about the leading diagonal, only six of these components are independent [4,5]. In addition, the incompressibility equation (1.1) for large plastic flow, effectively reduces this strain tensor to five independent components. It follows from this result that, if compatible plastic deformation is to occur in the individual crystals, throughout a polycrystalline aggregate, each metallic crystal must have at least five independent slip systems. This plastic compatibility condition is readily satisfied in both FCC and BCC metals, but not in HCP metals where additional twinning mechanisms are generally required for plastic flow processes to occur [6]. This is the explanation for the ability of FCC and BCC metals to undergo large plastic strains without fracture, and the inability of HCP metals to exhibit the same degree of ductility.

When plastic deformation occurs in a single crystal the yield shear stress τ_{sc} to move dislocations across the active slip plane is related to the equivalent force f_e per unit length of dislocation and the lattice Burger's vector b by the equation:

$$\tau_{sc} = f_e/b \; ; \qquad (1.2)$$

where f_e can be regarded as the summation of the effects of Peierls-Nabarro lattice resistance, solid-solution strengthening and work hardening. The tensile stress σ_{sc} required to bring about plastic deformation in a single crystal obeying equation (1.2) can be given in terms of the orientation φ of the slip plane, the orientation λ of the slip direction (Fig. 1.1) and the yield shear stress τ_{sc} on the slip plane by:

$$\sigma_{sc} = \frac{\tau_{sc}}{\cos\varphi \cos\lambda} \qquad (1.3)$$

where $\cos\varphi \cos\lambda$ is the Schmid factor. If the plastic shear strain $\Delta\gamma_{sc}$ is produced by the shear stress τ_{sc} on the active slip plane, the corresponding tensile strain $\Delta\epsilon_{sc}$ in the single crystal is given approximately by:

$$\Delta\epsilon_{sc} = \cos\varphi \, \cos\lambda \, \Delta\gamma_{sc}. \qquad (1.4)$$

Hence, for a single crystal there are simple approximate relationships between the macroscopic stress σ_{sc} and plastic strain-increment $\Delta\epsilon_{sc}$, and the shear stress τ_{sc}

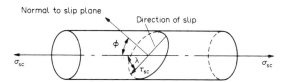

Fig. 1.1.   Cylindrical single crystal loaded in uniaxial tension.

and shear strain-increment $\Delta\gamma_{sc}$ on the crystallographic slip planes.

For the case of a polycrystalline aggregate the relationship between the macroscopic stresses and strains and the individual crystal properties are much more complex, and conditions of equilibrium and compatibility between individual crystals must be satisfied as plastic flow proceeds. The results from some recent models from the physical theory of plasticity, which seek to solve this problem of relating macroscopic and microscopic properties of polycrystalline aggregates, will be considered in Section 1.3. However, early work by Taylor [7] for polycrystals with more than five independent slip systems, which accounted in an approximate way for mutual grain boundary interactions, led to the following relation for the polycrystalline flow stress σ_{pc}:

$$\sigma_{pc} = M\,\tau_{sc} \; ; \tag{1.5}$$

where M is the Taylor factor. Both theoretical and experimental studies give a Taylor factor M ranging from 2 to 3 for the BCC and FCC metals [2,8].

In general, all elastic/plastic constitutive relationships for polycrystalline metals derived from the physical theory of plasticity, are much too complex to be of any great value in solving analytical problems involving elastic/plastic deformation of solids [9]; it is therefore necessary to base all analytical work on a mathematical model of an elastic/plastic continuum. With this approach the 'structureless' elastic continuum is assumed to exhibit plastic deformation only when the total stress state reaches a certain critical level which is represented by a 'yield criterion' [4,5]. The yield criterion is related to a given material through a parameter representing the experimentally determined plastic-yield stress, and in addition the yield criterion is used as a 'plastic potential' in deriving the plastic part of the elastic/plastic constitutive equations [4,5].

The form of the functional relation for a plastic-yield criterion, in an isotropic elastic/plastic continuum, can be derived in terms of the components of the stress tensor σ_{ij} as follows. Since a plastic-yield criterion must be independent of the particular reference axis system of σ_{ij}, if can always be written in terms of invariant functions of the three principal stresses σ_i. Now the principal stresses σ_i are the roots of the stress cubic equation:

$$\sigma^3 - I_1\sigma^2 - I_2\sigma - I_3 = 0 ; \qquad (1.6)$$

where I_1, I_2 and I_3 are the stress invariants defined by:

$$I_1 = \sigma_1 + \sigma_2 + \sigma_3 = \sigma_{ii} , \qquad (1.7a)$$

$$I_2 = -(\sigma_1\sigma_2 + \sigma_2\sigma_3 + \sigma_1\sigma_3) , \qquad (1.7b)$$

$$I_3 = \sigma_1\sigma_2\sigma_3 \qquad (1.7c)$$

Hence the principal stresses σ_i of the stress tensor σ_{ij} are represented unambiguously by the stress invariants I_1, I_2, I_3 and the plastic yield criterion must therefore have the form:

$$f(I_1, I_2, I_3) = 0 . \qquad (1.8)$$

In this most general form involving I_1 the yield criterion can be a function of the mean normal stress σ_m, since this is related to I_1 by (1.7a), i.e.:

$$\sigma_m = \tfrac{1}{3}\sigma_{ii} = \tfrac{1}{3}I_1 . \qquad (1.9)$$

However, in continuous polycrystalline aggregates the movement of dislocations along the active slip systems is controlled by the shear stress components τ_{sc} (equation (1.2)), and variations in the mean-normal stress σ_m have virtually no influence on the magnitude of the slip-plane shear stress. This follows from the fact that σ_m is a purely dilational stress with zero resolved shear stress on *all* possible orientations of the slip systems; the resolved shear stresses on active slip systems are in fact entirely determined by the deviatoric stresses, Fig. 1.2.

It is clear therefore that the functional form of the yield criterion in equation (1.8) can be rewritten entirely in terms of the deviatoric components of the stress tensor. The deviatoric stress tensor S_{ij} is defined by:

$$S_{ij} = \sigma_{ij} - \sigma_m\delta_{ij} , \qquad (1.10)$$

where δ_{ij} is the Kronecker delta, and the principal stresses S_i are given by the roots of the deviatoric stress cubic:

$$S^3 - J_1S^2 - J_2S - J_3 = 0 ; \qquad (1.11)$$

where the deviatoric stress invariants have the form:

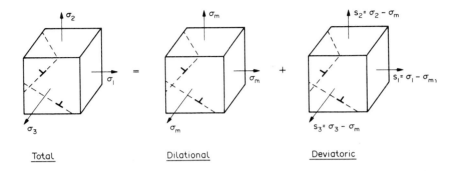

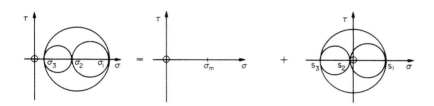

Fig. 1.2.   The total principal stress system (σ_1, σ_2, σ_3) and its decomposition into the dilational and deviatoric stress systems, with the corresponding Mohr's circles representations on the stress plane (σ, τ).

$$J_1 = S_1 + S_2 + S_3 = S_{ii} = 0, \qquad (1.12a)$$

$$J_2 = -(S_1 S_2 + S_2 S_3 + S_1 S_3) = \tfrac{1}{2}(S_{ij} S_{ij}), \qquad (1.12b)$$

$$J_3 = S_1 S_2 S_3 = \tfrac{1}{3} S_{ij} S_{jk} S_{ki}. \qquad (1.12c)$$

(The zero value of J_1 follows from equation (1.10)). The plastic yield criterion for continuous polycrystalline metals, deforming by crystallographic slip processes, must therefore have the form:

$$f(J_2, J_3) = 0. \qquad (1.13)$$

The most basic form of plastic yield criterion consistent with (1.13) is that due to von Mises [4,5] which is based on the assumption that J_3 has a negligible influence on plastic yielding, and thus (1.13) reduces to:

$$f(J_2) = 0 . \tag{1.14}$$

The particular form of $f(J_2)$ in the von Mises yield criterion is $(J_2 - k^2)$ where k is the yield shear stress of the plastic solid. Hence the von Mises yield criterion can be written in the following forms:

$$J_2 = \tfrac{1}{2} S_{ij} S_{ij} = \tfrac{1}{2} (S_1^2 + S_2^2 + S_3^2) =$$

$$\tfrac{1}{6} \left[(\sigma_1 - \sigma_2)^2 + (\sigma_2 - \sigma_3)^2 + (\sigma_1 - \sigma_3)^2 \right] = k^2 ; \tag{1.15}$$

and this can be rewritten in a more compact form for use in subsequent work as:

$$(\sigma_1 - \sigma_2)^2 + (\sigma_1 - \sigma_3)^2 + (\sigma_2 - \sigma_3)^2 = 2Y^2 ; \tag{1.16}$$

where Y is the uniaxial yield stress. This equation now gives the critical combination of principal stresses $(\sigma_1, \sigma_2, \sigma_3)$ which will bring about a state of incipient plastic deformation in a particular metal, having a uniaxial yield stress Y measured in a laboratory experiment.

1.2   The Concept of a Yield Surface and an Associated Flow Law in the Mathematical Theory of Plasticity.

One of the main advantages of writing the von Mises yield criterion (1.16) in terms of the three principal stresses (σ_i) of the stress tensor σ_{ij} is that the critical condition for incipient plastic deformation can be represented on a three-dimensional graph in principal-stress space, where the principal stresses $(\sigma_1, \sigma_2, \sigma_3)$ form the cartesian coordinates, Fig. 1.3. (If the yield criterion had been written in terms of the general components of the stress tensor σ_{ij}, a nine-dimensional space would be needed to represent the incipient plastic state.) It is readily seen from the form of equation (1.16) that the von Mises yield criterion is represented by a circular cylindrical surface in principal-stress space, with the cylinder axis coinciding with the line $\sigma_1 = \sigma_2 = \sigma_3$ and lying normal to the π plane $(\sigma_1 + \sigma_2 + \sigma_3) = 0$, Fig. 1.3. The plastic yield criterion is therefore equivalent to a yield surface in stress space. For all points $(\sigma_1, \sigma_2, \sigma_3)$ lying inside the yield cylinder, the solid has a purely elastic response and plastic deformation only initiates when the stress vector representing $(\sigma_1, \sigma_2, \sigma_3)$ touches the yield surface, Fig. 1.3.

The cylindrical form of the yield surface is a direct consequence of the fact that variations in the dilational or mean-normal stress $\sigma_m = \tfrac{1}{3}(\sigma_1 + \sigma_2 + \sigma_3)$ have no influence on the magnitudes of the deviatoric stresses and therefore on J_2, Fig. 1.2. From the geometry of Fig. 1.3 it is clear that the π plane $(\sigma_1 + \sigma_2 + \sigma_3 = 0)$ represents zero mean-normal stress σ_m and all points along the cylindrical axis

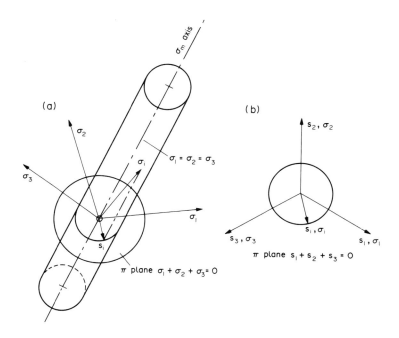

Fig. 1.3. (a) The von Mises yield surface $f(J_2) = 0$ in principal-stress
space $(\sigma_1, \sigma_2, \sigma_3)$.
(b) The projection of the cylindrical yield surface $f(J_2) = 0$ onto
the π-plane $(\sigma_1 + \sigma_2 + \sigma_3) = (S_1 + S_2 + S_3) = 0$.

$(\sigma_1 = \sigma_2 = \sigma_3)$ represent only variations in mean-normal stress. It follows from
this 'uncoupling' of the dilational and deviatoric stresses, in principal-stress space,
that the cylindrical yield surface can be represented entirely in terms of its projection onto the π plane, which is equivalent to replacing the principal stresses $(\sigma_1, \sigma_2,$
$\sigma_3)$ by the deviatoric principal stresses (S_1, S_2, S_3), from equation (1.10), Fig. 1.3(b).
Thus the yield surface can be transformed into a circular yield locus on the π plane
$(S_1 + S_2 + S_3 = 0)$, with three coordinate axes (S_1, S_2, S_3) displaced
by 120° and all coordinates reduced by a factor of $\sqrt{3/2}$ in the vertical projection. The
π plane projection of yield surfaces allows a relatively simple representation of
plastic-flow laws in the mathematical theory of plasticity and is also useful in giving
a concise format for presenting experimental results.

When a solid is subjected to a stress system σ_{ij}, in which the corresponding stress vector remains inside the yield surface, only elastic strains occur and any elastic strain increments $d\epsilon_{ij}^e$ are governed by Hookes' law. On the other hand, when the active stress vector σ_{ij} touches the yield surface both elastic- and plastic-strain increments $(d\epsilon_{ij}^p)$ occur simultaneously to give a combined strain-increment $(d\epsilon_{ij}^e + d\epsilon_{ij}^p)$. The constitutive equations for the plastic strain-increments $d\epsilon_{ij}^p$ can be derived from the geometry of the yield surface, which acts as a plastic potential, with the aid of Drucker's posulate [10] and the condition for continuity of elastic/ plastic deformation [4,5].

We now continue the discussion in terms of three-dimensional principal stress space, rather than in terms of the nine-dimensional space which is required to represent the general stress σ_{ij} and strain-increment $d\epsilon_{ij}^p$ tensors. The results therefore apply strictly only to cases of plastic flow where the principal axes remain fixed relative to an element. However, all the following results apply in principle to the general nine-dimensional space [4,5] and the general equations are obtained from the following results by replacing σ_i by σ_{ij} and $d\epsilon_i^p$ by $d\epsilon_{ij}^p$.

The plastic strain-increments $d\epsilon_i^p$ can be represented by a vector in principal stress space, if it is assumed that the strain-increment components are multiplied by an appropriate material parameter (e.g., the elastic shear modulus G) to give units of stress. Now it is readily proved that this plastic strain-increment vector will always lie in the π plane (Fig 1.3.), and this follows directly from a comparison of the incompressibility condition for plastic flow by crystallographic slip processes, equation (1.1), and the equation to the π plane:

$$\sigma_1 + \sigma_2 + \sigma_3 = S_1 + S_2 + S_3 = 0. \qquad (1.17)$$

Hence, the laws of plastic flow can be deduced entirely from the geometry of the π plane representation of the yield surface, Fig. 1.3(b).

In general, plastic flow is accompanied by a work-hardening effect which results in a change in the geometry of the yield surface [3-5]. In real materials these geometrical changes to the shape and location of the yield surface can be quite large and complex, even with small amounts of plastic deformation, and these general work-hardening effects will be considered in more detail in the following section. However, in establishing the laws of plastic flow, it is sufficient to assume a simple isotropic hardening effect in which the yield surface expands into an adjacent concentric circle, Fig. 1.4, and this will be called a 'loading surface' to distinguish it from the initial yield surface. This type of loading surface, of course, neglects the well-known Bauschinger effect [4,5].

In order to derive the laws of plastic flow we first note the *continuity condition* [4,5] which requires that all stress increments lying tangential to the existing yield locus (i.e., neutral increments) are accompanied by zero plastic strain-increment. This is

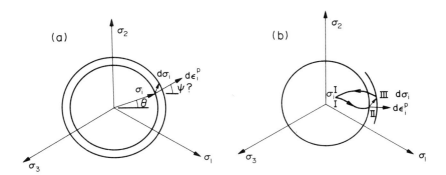

Fig. 1.4. (a) The π-plane projections of the active stress vector σ_i, the stress-increment vector $d\sigma_i$ and the currently undetermined orientation ψ of the plastic strain-increment vector $d\epsilon_i^p$, for a plastic solid with isotropic hardening.
(b) An arbitrary closed path (I → II → III → I) beginning at the initial stress σ_i^I, within the yield surface, and involving plastic flow between II and III.

the fundamental condition that ensures continuity of elastic/plastic strains as the direction of the stress-increment vector changes continuously from outside to inside the yield locus. The continuity condition is written in terms of the yield function f in the following form:

$$df = \frac{\partial f}{\partial \sigma_i} \, d\sigma_i = 0 .$$

$$(1.18)$$

We now consider Drucker's postulate [10] for the incremental stability of an elastic/plastic work-hardening solid, which can be stated as follows. For an element subjected to an initial state of stress σ_i^I, any quasi-static loading cycle I→II→III→I (Fig 1.4(b)) by *additional* stresses ($\sigma_i - \sigma_i^I$), will always cause the additional stresses to do positive work if a plastic strain-increment $d\epsilon_i^p$ occurs. Since the plastic strain-increment over the path II-III is infinitesimal and the work done on the elastic strains in a closed path is zero, we obtain the following inequality characterising stable plastic flow:

$$(\sigma_i - \sigma_i^I) \, d\epsilon_i^p > 0 .$$

$$(1.19)$$

If the initial stress state is now taken to be a point II (Fig. 1.4(b)) a closed loading and unloading cycle II→III→II leads to the following fundamental inequality of stable plastic flow:

$$d\sigma_i \, d\epsilon_i^p > 0 .$$

$$(1.20)$$

Under conditions of ideal non-hardening plasticity this inequality would be replaced by the equality $d\sigma_i \, d\epsilon_i^p = 0$. It should be noted at this point that Drucker's quasi-thermodynamic postulate for stable plastic flow is much stronger than a simple thermodynamic principle which requires only that the work done by the *total* stresses should be non-negative.

It is readily shown, with the aid of counter examples [5,11], that the continuity condition equation (1.18) and the stability inequalities in equations (1.19) and (1.20) establish the necessity for the plastic strain-increment vector $d\epsilon_i^p$ to lie normal to the yield or loading surface, and for yield and loading surfaces to be convex. These results follow from the fact that equation (1.19) requires the scalar product of $(\sigma_i - \sigma_i^l)$ and $d\epsilon_i^p$ to be always positive and equation (1.18) requires that zero plastic strain increments accompany neutral increments of stress. Hence, for a given active stress vector σ_{ij} on an arbitrary yield surface of convex form, the corresponding plastic strain-increment vector $d\epsilon_{ij}^p$ will always lie normal to the tangent plane at the point σ_{ij}, Fig. 1.5(a). The inequalities in equations (1.19) and (1.20) are of fundamental importance in the mathematical theory of plasticity and form the basis of the extremum principles which lead to upper and lower bound theorems for estimating the loads in plastic flow processes [4,5].

These results establish the necessity for an associated flow law in plasticity relating the plastic strain-increments $d\epsilon_i^p$ to the yield or loading function f. Now the continuity condition (df = 0) and the normality rule, together, require the $d\epsilon_i^p$ be proportional to the product of df and the direction cosines of the normal to f (i.e., $\partial f/\partial \sigma_i$), hence the associated flow rule for plastic strain-increments has the form:

$$d\epsilon_i^p = h \frac{\partial f}{\partial \sigma_i} \, df \, , \text{ whenever } df \geq 0 \, ; \tag{1.20a}$$

$$d\epsilon_i^p = 0 \quad \text{when } df < 0 \text{ (i.e., unloading)} \, ; \tag{1.20b}$$

where h is a scalar factor of proportionality related to the work-hardening effect. For all conditions involving plastic loading, equation (1.20a) can be rewritten in the form:

$$d\epsilon_i^p = d\lambda \, \frac{\partial f}{\partial \sigma_i} \, ; \tag{1.21}$$

where $d\lambda$ is a scalar factor of proportionality related to the increment of plastic work and therefore depends on the complete history of plastic flow. For the case of the von Mises yield function $f(J_2) = 0$ (equations (1.14) and (1.16)) the term $\partial f/\partial \sigma_i$ in equation (1.21) is equal to the deviatoric principal stress S_i and equation (1.21) can therefore be written in the form due to Prandtl and Reuss [4,5]:

$$d\epsilon_i^p = d\lambda \, S_i \, . \tag{1.22}$$

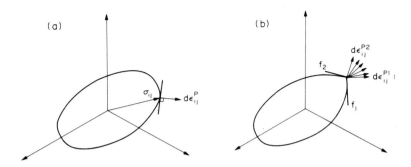

Fig. 1.5. (a) The normality rule of plastic flow, showing a smooth and convex yield surface with the uniquely determined direction of the associated plastic strain-increment $d\epsilon_{ij}^P$.
(b) The range of indeterminacy of $d\epsilon_{ij}^P$ at a yield vertex.

The constitutive equations for elastic/plastic deformation are therefore obtained in differential form by adding equation (1.22) to the elastic constitutive equations (Hooke's Law). The resulting equations for elastic/plastic deformation must generally be applied in differential form because the $d\lambda$ term in equation (1.22) is history dependent, unlike the purely-elastic constitutive equations which are independent of strain history.

The associated flow rule equation (1.22) implies that the principal axes of the plastic strain-increment always coincide with the principal axes of the current stress, and in general these differ from the principal axes of the *stress-increment*. The stress increments therefore determine only the *magnitude* of the plastic strain-increment vector $d\epsilon_i^P$, through the strain-history dependent parameter $d\lambda$, and not its direction. This theoretical result is intuitively consistent with dislocation processes of plastic flow in real polycrystalline aggregates, where the current stress state has brought about plastic flow by activating at least five independent slip systems in each individual crystal. Hence, in general, the application of a further non-proportional stress *increment* will tend to produce dislocation motion along the current set of active slip planes rather than activate a completely new set of slip planes.

An interesting consequence of the normality rule of plastic flow is that if a yield surface is not smooth at a certain point, Fig. 1.5(b), then the direction of the plastic strain increment vector $d\epsilon_{ij}^P$ is indeterminate between the angular limits set by the smooth surfaces adjacent to this point. A yield surface which contains such a singular point is described as containing a 'yield vertex' [5]. The question of whether yield vertices actually exist in real materials will be considered in the following section.

1.3 The Physical Theory of Plasticity : Plastic Work Hardening Effects and the Possible Existence of Yield Vertices

It is clear from the previous discussion that plastic flow processes in polycrystalline solids are extremely complex, and the mathematical continuum theories that have been developed to describe the constitutive relationship between the stresses and plastic strain-increments need to be applied with care when attempting to model the behaviour of real materials. In the mathematical theory of plasticity the yield criterion and associated flow rule are effectively developed for a structureless solid continuum, it is therefore useful to discuss the validity of the results in terms of results from an alternative 'physical' theory of plasticity. In the physical theory of plasticity the macroscopic behaviour of a plastic body is represented in terms of a mathematical model of a polycrystalline aggregate of cubic unit cells which individually deform by mechanisms analogous to the slip processes along crystallographic glide planes.

Some of the most useful and instructive work in the physical theory of plasticity was carried out by T.H. Lin and M. Ito [12-14], who developed two-dimensional and three-dimensional models for aggregates of cubic crystals with both FCC and HCP slip responses. The Lin and Ito results show firstly the very important sensitivity of the shape and location of a yield or loading surface to the *magnitude* of the additional plastic strain-increment used to define the yield stress. For example, they found that the initial-yield surface for an *infinitesimal* plastic strain-increment was close to a critical maximum shear-stress (Tresca) criterion [4,5]. On the other hand, if the equivalent plastic strain-increment $\Delta\bar{\epsilon}^p$ used to define the yield stress was of the order $\Delta\bar{\epsilon}^p = 0.1\mu\epsilon$ (where $0.1\mu\epsilon = 10^{-7}$ in/in) the corresponding initial-yield surface was close to the von Mises critical J_2 form. This sensitivity of the shape and location of a yield surface to the magnitude of the plastic strain-increment used to define it, is well confirmed by experiment [15-17]. In practice, of course, the accuracy of strain-measuring techniques require the plastic strain-increments to be no lower than $\sim 5\mu\epsilon$ if reliable results are to be obtained [15,16]; this is very much greater than the value of $\Delta\bar{\epsilon}^p = 0.1\mu\epsilon$ necessary for critical J_2 yielding and compares with a typical *elastic* strain at the initial yield point of $\sim 10\mu\epsilon$. The result emphasises the need to apply caution in interpreting results from the physical theory of plasticity when the plastic strain-increments are infinitesimal; infinitesimal strains can display effects which do not persist when small but finite strains occur.

The results from the Lin and Ito model of polycrystalline plasticity also show that with finite monotonic plastic straining beyond the initial yield surface the subsequent yield or loading surface (for $\Delta\bar{\epsilon}^p = 0.1\mu\epsilon$) exhibits a work hardening response which lies somewhere between the two extremes of isotropic hardening (where the loading surfaces form expanding concentric circles) and kinematic hardening (where the loading surfaces of constant radius translate across the deviatoric plane). This theoretical result is in agreement with experimental results

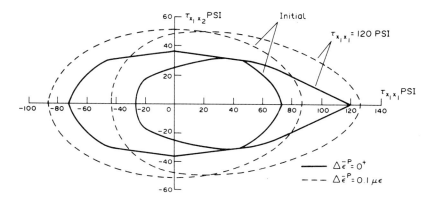

Fig. 1.6. Yield surfaces for a physical theory of plasticity, showing the initial yield surfaces defined by infinitesimal ($\Delta\bar\epsilon^P = 0^+$) and very small finite strain ($\Delta\bar\epsilon^P = 0.1\mu\epsilon$). Also showing the loading surfaces, following the application of a tensile stress $\tau_{11} = 120$ psi, defined by $\Delta\bar\epsilon^P = 0^+$ and $\Delta\bar\epsilon^P = 0.1\mu\epsilon$. (After T.H. Lin and M. Ito [13]).

[15,16] for yield and loading surfaces defined by a $5\mu\epsilon$ strain-increment; however, the same experimental results show that yield and loading surfaces display virtually isotropic hardening when measured by a relatively large plastic strain-increment $\sim 2000\mu\epsilon$.

Lin and Ito also used their physical-plasticity model to determine the current direction of the plastic strain-increment vector, on each yield and loading surface, and they found that for a loading surface defined by $\Delta\bar\epsilon^P = 2\mu\epsilon$ the strain-increment vector was within $3° \sim 6°$ of the direction of the normal to the tangent plane [13,14]; thus giving an approximate confirmation of the normality rule from the mathematical theory of plasticity.

Finally, the physical theory of plasticity gives results that throw light on the vexed question of the existence of yield vertices. Lin and Ito constructed the yield and loading surfaces, defined by an *infinitesimal* plastic strain-increment, for the case of monotonic loading to a tensile stress $\sim$ (1.67 x initial yield stress), and these results indicated the existence of a yield vertex at the corresponding point on the loading surface, Fig. 1.6. However, the same model showed that when the yield and loading surfaces were defined by a $0.1\mu\epsilon$ plastic strain-increment the yield vertex disappeared from the loading surface to leave it completely smooth, Fig. 1.6. Hence, a plastic strain-increment with a magnitude of only about 1% of the elastic strain at the initial yield point, is sufficient to completely eliminate yield-vertex effects in the physical theory of plasticity. It seems highly unlikely, therefore, that yield-vertex effects can have any influence on sustained processes of tensile,

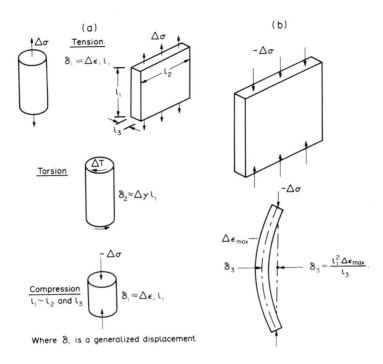

Fig. 1.7. (a) Modes of plastic deformation in which small plastic strains
always correspond to small geometrical displacements.
(b) The elastic/plastic buckling mode in a plate where small plastic strains can correspond to large transverse displacements.

torsional or compressive plastic flow, of the type that lead to failure by ductile fracture, where all geometrical displacement-increments accompanying small plastic
strain-increments remain small, Fig. 1.7(a). The highly transitory yield-vertex
phenomenon may, however, have some influence on problems involving the compressive elastic/plastic buckling of plates and shells [18], where large transverse
displacements can accompany very small through-thickness plastic bending
strains, Fig. 1.7(b).

A comprehensive survey of experimental studies of yield and loading surfaces,
and the possible existence of yield-vertex effects, has been carried out by Hecker
[16] and his survey shows no conclusive evidence to support the existence of a
yield-vertex effect in real materials. More recent theoretical work on the physical
theory of plasticity for a FCC polycrystalline model [19] also shows no evidence of
yield-vertex formation and at the same time confirms experimental results showing

work-hardening behaviour lying between the limits of isotropic and kinematic hardening.

It can therefore be concluded that the basic concepts of the mathematical theory of plasticity, including convexity of the yield and loading surfaces and the normality rule of plastic flow, are well supported both by the physical theory of plasticity and experimental results. In addition, both the physical theory of plasticity and experimental results confirm that the work-hardening effect in a polycrystalline aggregate lies somewhere between the two extremes of isotropic and kinematic hardening; with a tendency to approach the isotropic state with increase in the magnitude of the plastic strain-increments used to define yielding. Finally, experimental results show no convincing evidence of yield-vertex formation [16] and the physical theory of plasticity [14] suggests that if a yield vertex exists at all it is a highly transitory effect which disappears with a plastic strain-increment as small as $0.1\,\mu\epsilon$ (i.e., about 1% of the *elastic* yield-point strain). The question of the existence of yield vertices will be considered again, in a subsequent discussion of models of the ductile-fracture process, in Chapter 3.

1.4 The General Shape of Yield and Loading Surfaces in Three-Dimensional Principal-Stress Space

The classical mathematical theory of plasticity is based on the assumption that plastic yielding is a function only of the deviatoric stresses (i.e., $f(J_3, J_3) = 0$), and that subsequent work-hardening effects are isotropic. The yield surface and loading surface are therefore represented by a set of concentric cylinders related to the stress/strain curve as shown in the π-plane projection of Fig. 1.8. Since both the physical theory of plasticity [14, 19] and experimental results [15, 16] show a significant Bauschinger effect and a relatively small deviation from isotropic hardening towards kinematic hardening, a more precise representation of the yield and loading surfaces would be similar to the non-concentric 'cylinders' shown in the π-plane of Fig. 1.8(c). However, in most analytical problems involving the modelling of ductile-fracture processes, the active stress and plastic strain-increment vectors remain in the same small-angle ($\leqslant \pi/3$) sector of the π-plane. Hence, for all ductility models satisfying this condition, the yield and loading surfaces can be regarded as a set of concentric cylinders of increasing diameter, with axes of symmetry coinciding with the dilational (or mean-normal) stress σ_m axis, $\sigma_1 = \sigma_2 = \sigma_3$, Fig. 1.9(a).

Both the mathematical and physical theories of plasticity are based on the assumption that the polycrystalline solid is completely continuous, with no internal voids or surfaces, and that *all* points within the solid undergo plastic deformation by a dilationless process of crystallographic slip. This leads directly to yield and loading functions which are dependent only on the deviatoric stresses (i.e., $f(J_2, J_3) = 0$) and

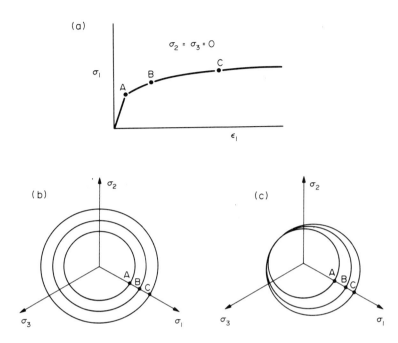

Fig. 1.8. The stress/strain curve in uniaxial tension (a) and its effect on the
yield and loading surfaces for (b) isotropic hardening and (c) hardening with a kinematic component.

thus have the form of cylinders in principal-stress space, Figs. 1.3 and 1.9(a),
aligned with the dilational-stress axis. If on the other hand the elastic/plastic solid
contains a distribution of internal voids, it is no longer possible to assume that plastic yielding will *always* be independent of the dilational or mean-normal stress σ_m,
since the presence of internal voids might lead to a truncation of the yield or loading cylinders at sufficiently high values of σ_m, Fig. 1.9(b). In this case the yield surfaces would have the more general form $f(I_1, I_2, I_3) = 0$, from equation (1.8). The
way in which microvoids are formed in real materials, under sustained plastic flow,
is discussed in the following section; the truncating effect of microvoids on the
yield and loading surfaces is considered in detail in Chapters 2 and 3.

1.5 Sustained Plastic Flow in the Presence of Microstructural Second-Phase Particles and Inclusions

Experimental results conclusively show [20-22] that in the initial stages of plastic
flow, in metals and alloys, inclusions and second-phase particles act as sites for the

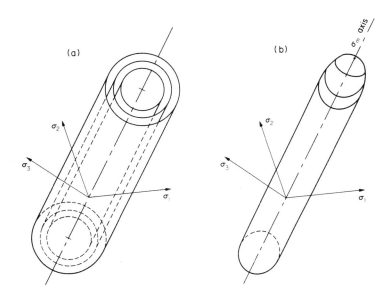

Fig. 1.9. (a) The yield and loading cylinders for the case of an assumed
             isotropic hardening effect.
          (b) Possible truncation of the yield cylinder at high values of σ_m
             for a solid containing a volume-fraction of microvoids.

nucleation of microvoids. In general, the extent of microstructural damage
increases with increasing plastic strain [22-25], and it is often found that the larger
second-phase particles and inclusions exhibit damage at smaller strains than the
smaller particles and inclusions [22]. There is also a tendency for the larger particles to fail by cracking and the smaller particles to fail by decohesion of the particle/
matrix interface [22]. Typical results for manganese sulphide inclusions in a low
alloy steel [26] and cementite particles in a 1% carbon steel [23] are shown in
Fig. 1.10, where the degree of microstructural damage is plotted against the total
plastic strain. There are, however, many complicating secondary effects acting in
parallel with the general effects of strain-related microstructural damage and a
complete review of these aspects of the microvoid nucleation problem is given by
Van Stone et al. [22]; including the relatively rare cases where microvoids are nucleated in the absence of particles at blocked slip-bands.

Following the nucleation of microvoids, at the sites of damaged particles and inclusions, the externally applied stress and plastic strain-rate field will lead to the continuous plastic growth of the microvoids. Numerous models have been presented
in attempts to describe the plastic growth of microvoids in plastic-flow fields, but

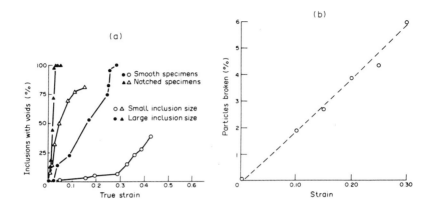

Fig. 1.10. (a) The effect of tensile strain on the nucleation of microvoids
in a AISI 4340 steel of both high purity (small inclusion size)
and commercial purity (large inclusion size). (After T.B. Cox
and J.R. Low, Metallurgical Transactions, 1974,5, p.1457).
(b) Proportion of broken carbide particles with increasing tensile strain in a 1% carbon steel. After J. Gurland [23].

by far the most satisfactory model for obtaining the continuous volumetric growth
and shape change of individual microvoids is that presented by Rice and Tracey
[27]. In the following work the Rice-Tracey equations will be used throughout to
model the plastic growth of microvoids with increasing plastic strain. It is readily
shown by an integration of the Rice-Tracey equations that, except at very high
mean-normal stress levels, the growth of microvoids should be predominantly an
extensional growth in the direction of the maximum principal stress accompanied
by only a relatively small dilational or volumetric growth [22,28,29]. This result is
in general quantitative agreement with experimental studies of void growth in the
central regions of necked tension specimens [25,30,31] where the maximum
mean-normal stress $\sigma_m \sim 0.7Y$, and microvoids grow predominantly by axial
extension. For plastic flow processes at higher mean-normal stress levels $\sigma_m \sim$
1.5Y (e.g., at the root of cracks and notches) the microvoids will also display substantial volumetric or dilational growth [22,29,32].

The plastic growth of microvoids continues with increasing plastic strain until a
critical condition is reached for the localised plastic instability or 'internal necking'
of the intervoid matrix, across a sheet of microvoids [20,22,28,33-35]. The controlling effect of microvoid populations on ductile-fracture strains is confirmed by

extensive experimental results [22-26] which show an inverse relationship between fracture strain ϵ^F and volume fraction V_f of particles or voids; this effect is illustrated in Fig. 1.11 by the classical results of Edelson and Baldwin [36] for copper compacts of varying inclusion or void content.

1.6 Physical Mechanisms of Ductile Fracture in the Presence of Microvoids.

In BCC and FCC polycrystalline metals of very high purity, the absence of inclusions and second-phase particles leads virtually to a complete absence of microvoids in any sustained process of plastic flow. Under these conditions a uniaxial tension specimen will usually fail by virtually 100% reduction in area of the external neck that develops when the specimen becomes plastically unstable [22,33]. However, all engineering metals and alloys contain inclusions and second-phase particles, to a greater or lesser extent, and this leads to microvoid nucleation and growth, which is terminated at a very much earlier stage in a tension test by a localised *internal* necking of the intervoid matrix across a sheet of microvoids [20-22]. Under these conditions ductile fracture occurs in tension specimens at approximately 70% reduction in area of the external neck, by the classical 'cup and cone' fracture mechanism. The general effect of an increasing volume fraction V_f of inclusions or second-phase particles is to reduce both the true fracture strain $\epsilon^F = \ln(A_o/A_f)$ and the % reduction is area $RA = 100(A_o - A_f)/A_o$; where A_o and A_f are the original and final cross-sectional areas of the specimen-neck region. These general effects are illustrated by Edelson and Baldwin's results in Fig. 1.11 where an increase in V_f from 0.01 to 0.1 leads to a reduction in ϵ^F from 1.5 to ~ 0.2; a further increase in V_f to 0.26 reduces ϵ^F to ~ 0.05.

A clear understanding of the physical mechanisms of ductile fracture can be obtained from scanning-electron microscope (SEM) studies of the plastic flow and fracture of uniaxial tension specimens, and in the final part of this chapter we consider a specific experimental study of ductile fracture in tension specimens from a low-carbon steel of 0.22% carbon and 0.68% manganese content supplied to En 3 specification (B.S. 970-1955).

The En 3 tensile specimens were given a carbide dispersion and spheroidisation treatment, before testing, which consisted of austenitizing for 4 hrs at 900°C, followed by water quenching and tempering for 2 hrs at 600°C. The heat treatments being carried out in inert-atmosphere furnaces and the tempering process was followed by water quenching to avoid temper embrittlement. The resulting microstructure of the En 3 specimens was found to consist of a uniform dispersion of spheroidal carbide (Fe_3C) particles of the order of 0.5μm diameter, along with relatively large and widely separated manganese sulphide and alumina (Al_2O_3) inclusions.

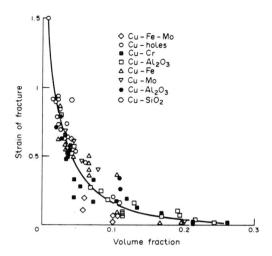

Fig. 1.11. The effect of volume fractions V_f of inclusions or voids in copper specimens on the true tensile strain at fracture ϵ^F. (After B.I. Edelson and W.M. Baldwin, Trans. ASM, ASM International, 1962, Vol.55, p.230).

The tensile specimens were tested under quasi-static conditions to give 'cup and cone' ductile-fracture surfaces with a reduction in area RA = 76%. This corresponds to a true fracture strain in the neck of $\epsilon^F = 1.43$, and estimates of the axial stress σ_z and mean-normal stress σ_m from Bridgman's equations [37] (c.f. Chapter 6.2) gave $\sigma_z = 1.46Y$ and $\sigma_m = 0.79Y$. The SEM fractographs of the relatively flat central region of the fracture surface are shown in Fig. 1.12 at various magnifications. At low magnification (X 20) the specimen displays the classical 'fibrous' appearance (Fig. 1.12(a)) and it is interesting to note that this specimen has an almost perfect 'cup' formation, with only a small part missing from the tip of the 'conical' fracture surface on the left-hand side of the figure. This latter 'double-cone' feature is of great significance when considering the mechanism of the so-called 'shear' fracture process, and will be considered in more detail below.

With increasing magnification of the central fibrous fracture-surface, to X 2K and X 5K (Figs. 1.12(b) and (c), respectively), the classical 'dimpled rupture' appearance is observed, and this results from the virtually complete necking-down of the inter-void matrix adjacent to damaged inclusions and carbide particles, giving the characteristic 'knife-edge' formations. A microsection on an axial plane, immediately adjacent to the central 'fibrous' fracture surface, is shown at magnifications of X 2.25K and X 5K in Figs. 1.13(a) and (b), respectively. The lower-magnification section of X 2.25K is a composite micrograph covering a relatively wide

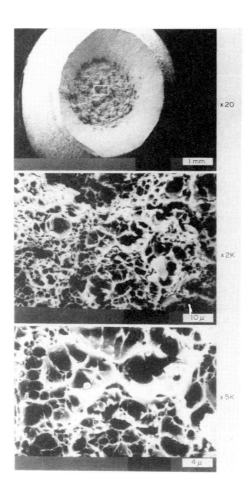

Fig. 1.12. Ductile-fracture surfaces in a low-carbon steel specimen, in the quenched and tempered condition, at varying magnifications: (a) X 20; (b) X 2K; (c) X 5K. The fractographs (b) and (c) are from the region of fibrous fracture, within the rectangular box shown in (a).

region adjacent to the fracture surface and this shows clear evidence of microvoid nucleation and growth from the sites of damaged manganese-sulphide and alumina inclusions Fig. 1.13(a). It is important to note, however, that any microvoids present in regions adjacent to a ductile-fracture surface are still relatively small and widely spaced [25,28-31], showing that ductile fracture is a process involving the sudden intervention of a plastically unstable mode of localised necking of the intervoid matrix. It is also interesting to note that the smaller and more uniformly dispersed carbide particles show no clear evidence of particle-cracking, and microvoid nucleation is therefore likely to be predominantly by decohesion of the matrix/particle bond which is less easily observed at these magnifications. This is confirmed by the microsection at X 5K in Fig 1.13(b) where only a relatively small proportion of the carbide particles are clearly seen to be associated with the nucleation of microvoids.

It is instructive to make comparisons between the fractographs in Fig. 1.12 and the adjacent microsections in Fig. 1.13, at similar magnifications. Comparing Fig. 1.12(b) with Fig. 1.13(a) and Fig. 1.12(c) with Fig. 1.13(b), it is clear that the larger and more widely spaced dimples correspond to the microvoids nucleated at inclusions, whereas, the small dimples correspond closely both in abundance and spacing to the carbide particles. There are two possible explanations for this apparently paradoxical situation in which microvoid coalescence occurs from what appear to be strongly-bonded inclusions. It is possible that the large and widely-spaced microvoids, nucleated at the site of inclusions, have the controlling influence on the conditions for incipient ductile fracture by localised internal necking of the intervoid matrix. Under these conditions the carbide particles within the regions of incipient localised necking will undergo an intense *additional* plastic straining sufficient to nucleate microvoids at virtually all of the locally enclosed particles; thus terminating the primary process of internal necking between inclusion-nucleated voids by a secondary process between carbide-nucleated voids, Fig. 1.13(c). The second alternative explanation for the apparent absence of microvoids at the site of carbide particles, is based on the possibility of either incomplete particle decohesion with undetectable void growth or spontaneous microvoid nucleation and coalescence [38] when the plastic stress field has been elevated to levels sufficiently close to the matrix/particle cohesive-bond strength. The relative effects of large and small microvoids on the critical conditions for ductile fracture and the apparent non-sequential nature of some void nucleation and growth processes [20,39] will be considered in detail in Chapters 2 to 4.

Experiments on tensile-test specimens, that are unloaded just prior to the final fracture, show that the central flat fibrous-fracture surface (Fig. 1.12(a)) develops in advance of the final catastrophic 'shear' fracture along the conical surfaces of the neck [40]. Hence, the so-called 'shear' fracture only forms *after* the initiation of the primary central fibrous fracture, and the conical 'shear' lips are therefore secondary fracture surfaces formed by the catastrophic propagation of the central 'fibrous' crack.

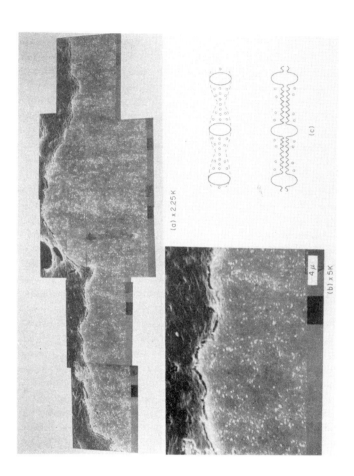

Fig. 1.13. Microsections of regions adjacent to the central fibrous-fracture surface of the specimen shown in Fig. 1.12, at magnifications of: (a) x 2.25K; (b) x 5K. Schematic diagram in (c) shows a possible mechanism of primary and secondary internal necking from large inclusion-nucleated voids and carbide particles, respectively.

However, although the 'shear' lips are secondary fracture effects in uniaxial tension tests, they are similar to the primary fracture processes that can develop at the
roots of grooves and notches. We will therefore consider the physical process of
'shear'-fracture in some detail at this stage and show that it is *not* generally a
'shear' fracture in the true sense of that description but is in fact a tensile fracture
on an inclined surface.

The SEM fractographs of the conical 'shear' fracture surface, for the En 3 specimen,
are shown at various magnifications in Fig. 1.14. At a magnification of X 20 the
'shear' lip displays a smoother appearance than the central fibrous fracture. However, at a magnification of X 500 (Fig. 1.14(b)) the surface is clearly of a fibrous
nature, with widely-separated large microvoids nucleated at the site of inclusions.
At a magnification of X 5K (Fig. 1.14(c)) the classical dimpled-rupture appearance
is observed, involving both inclusion- and carbide-nucleated microvoids, similar to
the central fibrous-fracture surface, Fig. 1.12(c). The primary difference between
these two fracture surfaces is the obvious tendency for the void-coalescence displacement vector to lie at approximately 45° to the 'shear'-lip surface (Fig. 1.14(c))
and ~ 90° to the central fibrous surface (Fig. 1.12(c)). It is important to note however
that, although microvoid coalescence has occurred on the 'shear'-lip surface at
approximately 45° to the axis of the tension specimen, this is *not* a shear fracture
in the sense that the two sides of the conical fracture surface have at any stage
undergone a *purely tangential* relative displacement. In fact, a purely tangential
relative-displacement around the almost completely-formed conical 'shear' lips,
shown in Figs. 1.12 and 1.14, must clearly be excluded on the grounds that it would
violate compatibility conditions for diametral points of the cone, as pointed out by
Cottrell [33].

The fact that the conical 'shear' lips are not true *shear* fractures is confirmed by a
microsection (Fig. 1.15) through the interesting 'double' conical surfaces seen at
the lower edge of the almost-complete cup and cone surface in Fig. 1.14(a). In this
case the low-magnification (X 20) section (Fig. 1.15) shows that the 'shear' lip
initiates at approximately 45° to the tensile axis and, with increasing radial distance
from the axis, the angular orientation of the fracture surfaces changes continuously to approximately 55°. There is then a discontinuous change in fracture surface orientation to approximately 160° and with further increase in radius this
changes continuously to a 120° orientation adjacent to the external surface of the
neck, Fig. 1.15. These 'double cone' fracture surfaces clearly cannot lie along shear
bands and a compound microsection at a magnification of 1K, close to the apex of
the 'double cone' (Fig. 1.15), confirms the absence of intense shear bands and
shows the classical 'tensile' mechanism of ductile fracture by internal necking between adjacent microvoids. Hence, it is clear that the conical fracture surface at a
tension-specimen neck is generally the result of the outward catastrophic crack
propagation of the initial central ductile crack and the path of propagation is likely
to be strongly influenced by the local stress-concentration effect at the tip of the
internal ductile crack.

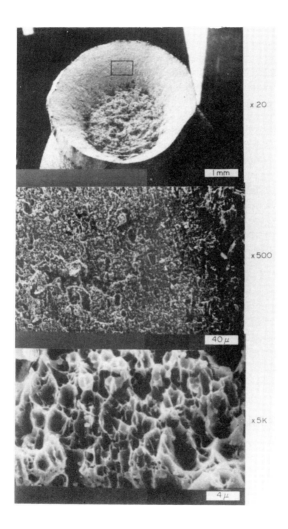

Fig. 1.14. Ductile fracture surfaces for the low-carbon steel specimen, showing the conical 'shear' fracture region at varying magnifications : (a) X 20; (b) X 500; (c) X 5K. The fractographs in (b) and (c) are from the region within the rectangular box shown in (a).

In relatively rare cases involving the ductile fracture of tension specimens of very low particle volume-fraction, such as commercially-pure copper, the central ductile crack may not immediately propagate to form a complete cup and cone fracture surface [40]. In this case a further small increment of plastic deformation may be required to bring about complete ductile fracture and this can lead to the development of a triangular-section plastic zone around the central crack which contains tangential velocity discontinuities on the boundaries; these tangential velocity-discontinuities produce intense shear bands along which the final conical fracture-surface then develops [40]. This mode of apparently true shear-fracture in discontinuous plastic-velocity fields is considered in detail in Chapters 6.5, 7.3 and 8.

In the subsequent chapters plastic continuum models are used to describe the various processes that bring about ductile fracture in metals and these are based on the understanding that relatively simple unit-cell models of microvoid nucleation, growth and coalescence can be applied to real ductile-fracture processes involving the simultaneous effects of the widely differing types and morphologies of inclusion and second-phase particles that can exist in a real metal. The basis for the application of these unit-cell models is the fact that a small ductile-fracture surface of area $\sim 0.25 mm^2$, similar to that indicated by the rectangle in the centre of Fig. 1.12(a), will contain approximately 10^5 individual microvoids of similar sizes and distributions to those shown in Fig. 1.12(c). There are therefore more than sufficient microvoids on an initial macroscopic fracture surface to justify the use of a unit-cell model representing the statistical average microvoid size and spacing for the particle or inclusion types that take on the controlling influence in any particular ductile-fracture process. In addition it should be noted that throughout the following work it is assumed that the plastic-continuum models are to be applied to real metals that have approximately homogeneous chemical and physical structures with no general tendency to pronounced segregation effects, particularly at grain boundaries. When prenounced microstructural segregation effects exist in a metal the following models can be applied in principle, but only on a localised quantitative basis.

An important aspect of the micro-modelling of ductile fracture, which follows directly from the above consideration of microstructural features, is that the ductile-fracture process must always be regarded as a mechanism of localised plastic instability occuring simultaneously in the intervoid matrix between *very large* numbers of coalescing microvoids which, by the statistical averaging effect, behave in the same way as an assembly of large numbers of unit cells. It is a failure to appreciate this point that has led to the appearance in the recent literature [41] of a number of rather loose assumptions about the nature of the ductile-fracture mechanism, based on observations from SEM micrographs showing a small number of closely-spaced and isolated microvoids in regions adjacent to an existing ductile-fracture surface. With this type of microvoid cluster it is possible for two

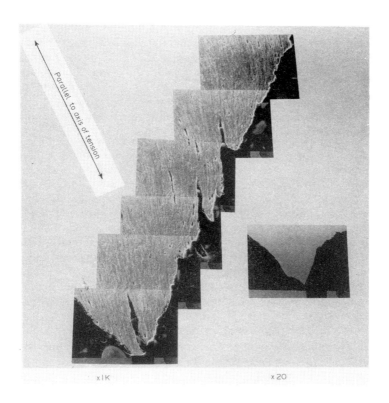

Fig. 1.15. Microsections of the conical 'shear' fracture region of the 'cup and cone' fracture surface, for the low-carbon steel specimen, at magnifications of: (a) X 20; (b) X 1K. The composite microsection in (b) shows the region adjacent to the apex of the 'double-cone' fracture surface.

adjacent microvoids to show an intervoid-matrix geometry in the form of a smoothly diminishing thin filament, which is unable to neck-down catastrophically due to the surrounding plastic constraint of the matrix on these isolated microvoids [41]. It is clear, therefore, that observations from micrographs of regions adjacent to an existing ductile-fracture surface must always be made with some caution, in the knowledge that any apparent mechanism of microvoid growth and coalescence must also be capable of existing between very large numbers of microvoids up to the point of catastrophic void coalescence. In other words any apparent fracture mechanism must be both statically and kinematically admissible in the presence of large numbers of microvoids, if it is to represent a real mechanism of ductile fracture. A further point to note when interpreting the SEM microsections of regions adjacent to ductile-fracture surfaces is the effect of the heavy etching that is necessary in the preparation of SEM microsections. When an isolated microvoid cluster is subjected to heavy etching, the intervoid matrix within the cluster is preferentially attacked and the resulting corrosion cavities can be easily misinterpreted as being part of the microvoid growth and coalescence process [41], for this reason SEM microsections of ductile-fracture specimens must always be interpreted with particular care.

REFERENCES

1. Cottrell, A.H. *The Mechanical Properties of Matter.* John Wiley, 1964.
2. Friedel, J. *Dislocations.* Pergamon Press, 1969.
3. Honeycombe, R.W.K. *The Plastic Deformation of Metals.*Arnold, 1969.
4. Hill, R. *The Mathematical Theory of Plasticity.* Clarendon Press, Oxford, 1950.
5. Kachanov, L.M. *Fundamentals of the Theory of Plasticity.* MIR Publishers, Moscow, 1974.
6. Kelly, A. and Macmillan, N.H. *Strong Solids.* 3rd Edition, Clarendon Press, Oxford., 1986.
7. Taylor, G.I. *J. Inst. Metals.* 1938, 62, p.307.
8. Hull, D. *Introduction to Dislocations* 2nd Edition, Pergamon Press, 1975.
9. Rice, J.R. *Constitutive Equations in Plasticity.* edited by A.S. Argon, MIT Press, Cambridge, MA, 1975, p.23.
10. Drucker, D.C. *Structural Mechanics.* edited by J.N. Goodier and N.J. Hoff, Pergamon Press, N.Y., 1960, p.407.
11. Hodge, P.G. Jr. *Continuum Mechanics* McGraw Hill, N.Y., 1970.
12. Lin, T.H. and Ito, M. *J. Mech. Phys. Solids,* 1965, 13, p.103.
13. Lin, T.H. and Ito, M. *Int. J. Engng. Sci.* 1966, 4, p.543.
14. Lin, T.H. *Advances in Applied Mechanics.* Vol. 11, edited by Chia-Shun Yih, Academic Press, N.Y., 1971, p.256.
15. Hecker, S.S. *Metallurgical Trans., 1973, 4, p.985.*
16. *Hecker, S.S. Constitutive Equations in Viscoplasticity: Computational and Engineering Aspects,* A.M.D., Vol. 20, edited by J.A. Stricklin and K.J. Saczalski, Am. Soc. Mech. Engrs., N.Y., 1976, p.1.
17. Hsu, T.C. *J. Strain Anal.* 1966, 1, p.331.
18. Sewell, M.J. *J. Mech. Phys. Solids,* 1974, 22, p.469.

19. Beradai, C., Berveiller, M., and Lipinski, P. *Int. J. Plasticity.* 1987, 3, p.143.
20. Wilsdorf, H.G.F. *Mat. Sci. Eng.* 1983, 59, p.1.
21. Leslie, W.C. *I.S.S. Trans.,* 1983, 2, p.1.
22. Van Stone, R.H., Cox, T.B., Low, J.R. Jr., and Prioda, J.A. *Int. Metals Reviews.* 1983, 30, p.157.
23. Gurland, J. *Acta Metall.* 1972, 20, p.735.
24. Greenfield, M.A., and Margolin, M. *Metallurgical Trans.* 1972, 3, p.2649.
25. Argon, A.S., and Im, J. *Metallurgical Trans.* 1975, 6, p.839
26. Cox, T.B., and Low, J.R. Jr. *Metallurgical, Trans.* 1974, 5, p.1457.
27. Rice, J.R., and Tracey, D.M. *J. Mech. Phys. Solids.* 1969, 17, p.201.
28. Thomason, P.F. *Prospects of Fracture Mechanics.* edited by G.C. Sih, H.C. Van Elst and D. Broek, Noordhoff, Netherlands, 1974, p.3.
29. Thomason, P.F. *Acta Metall.* 1981, 29, p.763.
30. Liu, C.T., and Gurland, J. *Trans. ASM,* 1968, 61, p.156.
31. Hayden, H.W., and Floreen, S. *Acta Metall.* 1969, 17, p.213.
32. Knott, J.F. *Met. Sci.* 1980, 14, p.327.
33. Cottrell, A.H. *Fracture.* edited by B.L. Averbach, Chapman and Hall, London, 1959, P.20.
34. Thomason, P.F. *J. Inst. Metals.* 1968, 96, p.360.
35. LeRoy, G., Embury, I.D., Edwards, G., and Ashby, M.F. *Acta Metall.* 1981, 29, p.1509.
36. Edelson, B.I., and Baldwin, W.M. *Trans. ASM,* 1962, 55, p.230.
37. Bridgman, P.W. *Trans. ASME.* 1944, 32, p.553.
38. Thomason, P.F. *Metal Sci. J.* 1969, 3, p.139.
39. Neumann, P. *Mat. Sci. Eng.* 1976, 25, p.217.
40. Bluhm, J.I., and Morrissey, R.J. *Proc. 1st Int. Conf. on Fracture.* Sendai, Japan, 1965, edited by T. Yokobori, T. Kawasaki and J.L. Swedlow, vol. 3, p.1739.
41. Park, I-G., and Thompson, A.W. *Acta Metall.* 1988, 36, p.1653.

CHAPTER 2

The Mechanics of Microvoid Nucleation
and Growth in Ductile Metals

2.1 Microvoid Nucleation at the Sites of Second-Phase Particles and Inclusions in a Ductile Plastic Matrix

Some of the most complex and difficult modelling problems in the theory of ductile fracture are concerned with the nucleation of microvoids at the sites of inclusions and second-phase particles in a plastically deforming matrix. The second-phase particles and inclusions can range in size from $\sim$ 0.01 μm, where dislocation models are required to describe the microvoid nucleation process [1,2], to $\geqslant$ 1μm where plastic continuum models are applicable [3,4]. In addition, the shape of the second-phase particles and inclusions can vary from a spheroidal form to a lamellar or irregular angular form [4,5]. These wide variations in particle size and shape normally exist simultaneously in a practical engineering alloy, and it should be emphasised that the complete history of deformation processing and machining from the 'as cast' state to the finish-machined engineering component cannot strictly be neglected when attempting to model the microvoid nucleation process in a particular alloy system. The microvoid nucleation problem in its entirety is therefore highly complex and a great deal of work remains to be done to obtain a complete understanding of the problem. In the following discussion we consider some recent progress in modelling the microvoid nucleation problem, and this is primarily concerned with the case of hard spheroidal particles which have a strong adhesive bond to the matrix.

A necessary condition for the nucleation of a microvoid, by decohesion of the particle/matrix interface, is that the elastic strain energy released by the particle is at least equal to the newly created surface energy [1-3,6]. However, this condition is by no means sufficient and is likely to be satisfied at the very onset of plastic deformation for particles in excess of $\sim$0.025 μm diameter [7]. It can therefore be concluded that a sufficient condition for microvoid nucleation by decohesion is the attainment of a critical normal stress σ_c at the particle/matrix interface [2,3].

30

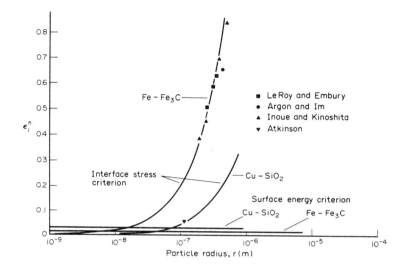

Fig. 2.1.   A comparison of the experimental microvoid-nucleation strains ϵ_1^n for various particle radii r, with the theoretical nucleation strains from the Brown and Stobbs model [1,2]. (After S.H. Goods and L.M. Brown [2]).

For the case of small spheroidal particles (radius r < 1 μm), where the local flow stress σ_l is dependent on the increasing rate of dislocation storage as the particle radius decreases, dislocation models [1,2] are required to estimate the total interface stress σ_T on a particle. It has been shown by Brown and Stobbs [1] that the dislocation density ρ_l around a particle is given approximately by $\rho_l = 1.7\, \epsilon_1/rb$ and thus the local flow stress is given by the expression:

$$\sigma_l \;=\; \alpha\mu b\,(\rho_l)^{1/2} \;=\; 1.3\alpha\mu\!\left(\frac{\epsilon_1 b}{r}\right)^{1/2} ; \qquad (2.1)$$

where μ is the shear modulus, b is the Burger's vector, ϵ_1 is the maximum macroscopic plastic strain and α is a constant lying between ~1/3 and 1/7. The presence of the particle imposes a local plastic constraint on the matrix thus elevating the local stress on the particle interface by a factor which has been estimated [1,2] to be ~4.2. Thus the elevated local stress on the particle interface is given by [2]:

$$\sigma_E \;=\; 5.4\alpha\mu\!\left(\frac{\epsilon_1 b}{r}\right)^{1/2} ; \qquad (2.2)$$

The maximum stress σ_T on the particle interface is therefore the sum of the elevated local stress σ_E and the macroscopic mean-normal σ_m and maximum deviatoric S_1 stresses:

$$\sigma_T = S_1 + \sigma_m + \sigma_E. \tag{2.3}$$

Hence the critical condition for microvoid nucleation by decohesion of the matrix/ particle interface is given by:

$$S_1 + \sigma_m + \sigma_E = \sigma_c; \tag{2.4}$$

where σ_c is the critical cohesive strength of the interface. By developing approximate expressions for the macroscopic flow stress $(S_1 + \sigma_m)$ it can be shown [1,2] on substituting eq. (2.2) into eq. (2.4) that the critical strain ϵ_1^n required to nucleate microvoids by particle decohesion is given by:

$$\epsilon_1^n = Kr(\sigma_c - \sigma_m)^2; \tag{2.5}$$

where K is a material constant related to the volume fraction of particles.

Equation (2.5) shows that for a given set of material and stress-field conditions the nucleation strain is a linear function of particle radius r, where $r \leqslant 1$ μm. Experiments on materials from the Cu-SiO_2 and Fe-Fe_3C systems [1,2] confirm the linear increase in microvoid nucleation strain ϵ_1^n with increase in particle radius (Fig. 2.1) and give nucleation strains varying from $\epsilon_1^n = 0.4$ at $r \approx 0.4$ μm to $\epsilon_1^n = 1.0$ at $r \approx 1$ μm. Further support for the validity of equation (2.5) is obtained [8] from experimental results for the nucleation strains at various mean-normal stress levels σ_m, for a number of spheroidized steel specimens having approximately the same volume fraction of carbide particles of similar radius. When these results are plotted on a graph of $\sqrt{\epsilon_1^n}$ against σ_m, Fig. 2.2, and fitted by the equivalent linear relationship from equation (2.5), an extrapolation to zero nucleation strain gives $\sigma_m = \sigma_c = 1200$ MPa. This value for σ_c is in good agreement with the result $\sigma_c = 1700$ MPa for carbides in steel obtained by Argon and Im [4], and is within the range of carbide-interface strengths for steels which has been estimated to be $1000 \sim 3000$ MPa.

The rate of dislocation storage around a particle, in a plastically deforming matrix, decreases with increase in particle size [1,2] and this results in the existence of a particle size above which the local strain-hardening effect (equation (2.1)) is of the same magnitude as that in the matrix. This critical particle size is of the order $r = 1$ μm and, for particles of larger size, continuum-plasticity models can be developed to describe the microvoid nucleation process in materials that do not exhibit coarse slip.

An approximate analysis by Argon et al. [3] of both the non-hardening and linear-hardening plastic flow around a circular-cylindrical inclusion, in a pure-shear

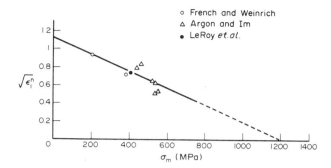

Fig. 2.2.   The critical microvoid nucleation strains ϵ_1^n for various mean-normal stresses σ_m in a spheroidised 1045 steel. The theoretical relation from the Brown and Stobbs model [1,8] is fitted to the results and projected to $\epsilon_1^n = 0$ to give $\sigma_c = 1200$ MPa for decohesion of the carbide/ferrite interface. (After G. Le Roy et al [8])

deformation field (Fig. 2.3a), shows that the maximum radial stress σ_r^{max} at the inclusion interface is of the order of 1.75 k (Fig. 2.3b); where k is the yield shear stress of the plastic matrix. Since the uniaxial yield stress $Y = \sqrt{3}k = 1.732$ k (from equation (1.15)), the maximum tensile stress on the inclusion interface can be represented by $\sigma_r^{max} \simeq Y = \bar{\sigma}$; where $\bar{\sigma}$ is the equivalent stress defined by [9]:

$$\bar{\sigma} = \left\{ \tfrac{1}{2} \left[(\sigma_1 - \sigma_2)^2 + (\sigma_1 - \sigma_3)^2 + (\sigma_2 - \sigma_3)^2 \right] \right\}^{\frac{1}{2}}. \qquad (2.6)$$

The mean normal stress σ_m is equal to zero for a state of pure shear (i.e., $\sigma_1 = k$, $\sigma_2 = -k$, $\sigma_3 = 0$) and it was therefore proposed [3] that a good approximation for the maximum stress σ_r^{max} at the inclusion/matrix interface in a general plastic-flow field could be obtained by adding $\bar{\sigma}$ and σ_m to give:

$$\sigma_r^{max} = \bar{\sigma} + \sigma_m \quad . \qquad (2.7)$$

The critical condition for microvoid nucleation, by decohesion of the particle/matrix interface, is reached when $\sigma_r^{max} = \sigma_c$ and is therefore represented by:

$$\bar{\sigma} + \sigma_m = \sigma_c. \qquad (2.8)$$

The primary difference between this criterion for microvoid nucleation and that for sub-micron sized particles [1,2] in equation (2.4), is that the parameters in equation (2.8) are all independent of particle radius. It should also be noted that although

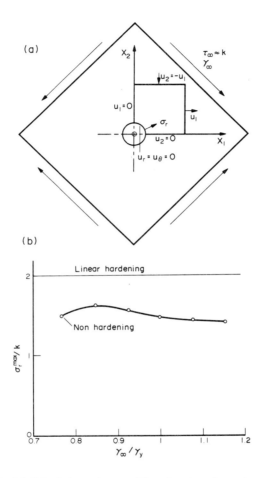

Fig. 2.3. (a) Velocity boundary conditions for a two-dimensional model
               of plastic flow around a hard circular inclusion.
           (b) Maximum radial stress σ_r^{max}/k on inclusion interface, with
               increasing remote shear strain γ_∞/γ_y, for a non-hardening
               matrix. Also showing the corresponding linear-hardening
               result. (After A.S. Argon, J. Im and R. Safoglu, Metallurgical
               Transactions, 1975, 6A, p.825).

equation (2.8) is a stress criterion it will be possible in any particular problem to establish a corresponding critical nucleation-strain, through the relationship of both $\bar{\sigma}$ and σ_m to the total plastic strain ϵ_1.

This approach was adopted by Argon and Im [4] in experiments on uniaxial tension specimens of both cylindrical and pre-necked form, which gave a wide range of σ_m values at the points along the various specimen axes where incipient microvoid decohesion occurred. For the case of a spheroidised medium-carbon steel, with approximately equiaxed carbides of up to 1 μm radius, it was found (Fig. 2.4a) that the critical decohesion strength of the carbide/matrix interface was virtually constant at a value of $\sigma_c \approx 1700$ MPa and this was achieved at nucleation strains in the range $0.27 \leqslant \epsilon_1^n \leqslant 0.67$ as the mean-normal stress varied in the range $0.47Y \leqslant \sigma_m \leqslant 0.78Y$. Similar effects were found [4] for a copper-chromium alloy (Fig. 2.4b), with Cu-Cr particles, and a maraging steel (Fig. 2.4c) with TiC particles. It was also noted by Argon and Im [4] that, in common with many previous experiments [5,10-12], there was clear evidence that the nucleation strain for larger particles was lower than that for smaller particles. This result does not appear to comply with the models presented above for submicron sized particles [1,2], where the nucleation strain from equation (2.5) shows a linear increase with particle radius, and for larger particles ($r \geq 1$ μm) [3,4] where the critical decohesion condition (equation (2.8)) is independent of particle radius. However, this discrepancy between theory and experiment can be explained by a number of complicating features of the microvoid nucleation problem.

It was suggested by Argon and Im [4] that the premature decohesion of large particles was the result of an interaction of the local plastic-strain fields of closely adjacent particles and they developed a continuum model for particle interaction in an attempt to rationalise their results. This model shows, however, that very large local volume-fractions of particles, $V_f = 0.3 \sim 0.5$, are needed to give any appreciable elevation of the particle interface stress by particle interaction. The premature decohesion of large particles is observed with very much lower particle volume fractions ($V_f \leqslant 0.1$) [10-12] and it is therefore necessary to consider alternative mechanisms to account for the observed effects.

The microvoid nucleation problem is greatly complicated by the wide range of particle types and particle morphologies that can exist in a commercially-produced alloy system and this can result in a variety of microvoid-nucleation mechanisms operating simultaneously. The following discussion of microvoid-nucleation effects is primarily concerned with the case of alloy steel systems where a wide range of carbide, oxide, sulphide and nitride particles can exist, including the predominant iron carbides (Fe_3C), aluminium oxides (Al_2O_3) and manganese sulphides (MnS_2); however, the void-nucleation mechanisms discussed below will apply in principle to all ductile alloy systems containing hard particles.

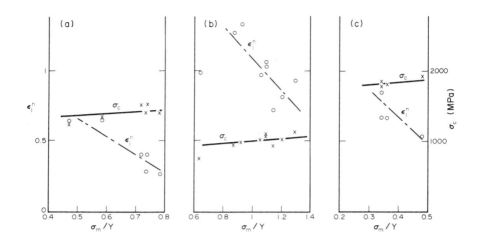

Fig. 2.4.   Variations of microvoid nucleation strain ϵ_1^n and critical inter-
face-decohesion stress σ_c , with mean-normal stress σ_m , for
1045 steel (a), Cu-0.6%Cr (b) and VM 300 maraging steel (c)
alloy/particle systems. (From experimental results of A.S.
Argon and J.Im, Metallurgical Transactions, 1975, 6A, p.839).

In practical alloy-steel systems aluminium oxide and manganese sulphide particles are always present, to a greater or lesser extent, and in the hot-working pro-
duction processes the hard oxide inclusions retain their spheroidal shape while the
softer manganese-sulphide inclusions become highly elongated in the direction of
hot-working. In addition, in high-carbon steels, a substantial proportion of the carbides are insoluble in the austenite matrix, at the hot-working temperature, and
these 'primary' carbides are subjected to severe damage and fragmentation in the
course of the hot-working process. Consequently, the larger oxides, sulphides and
primary carbides in steel systems are likely to be in a partially-damaged state at the
end of the steel-production process, and this might consist of partial debonding of
the particle/matrix interface or partial internal cracking of the hard particles. Hence,
in the case of steels, only the 'secondary' spheroidal carbides, precipitated from
the austenitic phase by quenching and tempering treatments, can be regarded
with any confidence as undamaged and strongly bonded to the matrix. It follows
from this that an adequate model of the microvoid nucleation process in steels
needs to be developed on the basis of at least two populations of particles in a
given material [8]. The first population N_1 would consist of the strong and firmly-
bonded 'secondary' carbides, to which the Brown and Stobbs dislocation model [1,2]
applies for submicron sized particles and the Argon et al. continuum model [3,4]
applies for larger particles. The second population N_2, which in some cases might

need to be further sub-divided, would consist of partially-damaged primary carbides and inclusions of the alumina and manganese-sulphide types.

The first population N_1 of firmly-bonded 'secondary' carbides would require total plastic strains in the range $0.3 \sim 1.0$ to nucleate microvoids by decohesion [1-4], and this could be well in excess of the nucleation strains for the second population N_2 of damaged 'primary' carbides and inclusions. Under these conditions the second population N_2 of partially-damaged particles could take over the controlling influence on microvoid nucleation and it is therefore interesting to speculate on how adequate models might be developed to describe the process. Leaving aside the statistical approach that would be a necessary part of any complete quantitative model, there are basically two mechanisms of void formation to be considered for the N_2 population.

The first void-nucleation mechanism applies to inclusions and second-phase particles which are elongated in the direction of maximum principal plastic-strain, where fibre-loading effects [13] can lead to the build-up of a large axial tensile stress (Fig. 2.5(a)) and premature cracking of the particle at relatively small plastic strains in the range $0.05 < \epsilon_1{}^p < 0.1$ [14]. The fibre-loading effect can occur in rod or plate-shaped carbides and manganese-sulphide inclusions in steels [14-16], although in the case of manganese sulphides the low interface bond strength may often result in decohesion of the particle/matrix interface [12].

The second void-nucleation mechanism applies to approximately equiaxed particles which are in a partially damaged state and contain very small crack-like defects, Fig. 2.5(b). Under these conditions the particle may bring about premature microvoid nucleation when the maximum tensile stress $\sigma_r{}^{max}$ on the particle interface is sufficient to cause complete cracking by catastrophic propagation of the internal defect [12,17,18]. This critical interface stress σ_f can be estimated approximately by the Griffith equation:

$$\sigma_f = A \sqrt{\frac{2E\gamma}{\pi c}} \tag{2.9}$$

where E is Young's modulus of elasticity, γ is the fracture surface energy, c is the internal crack length and A is a geometrical constant. Now the carbide and oxide particles present in steels have typical fracture surface energies γ in the range 0.01 to 0.05 KJm^{-2}, elastic moduli E from 170 to 510 GPa and crack-free fracture strengths σ_{th} from 3 to 7 GPa [19,20]. Hence, if the maximum tensile stress $\sim\sigma_r{}^{max}$ generated in the particles by the plastically-deforming ferrite matrix is in the range 1.4 to 1.9 GPa [4] the crack-free fracture strengths σ_{th} will never be reached, but any sufficiently-large particles containing microcracks of length c in the range 0.3 to 8 μm could nucleate microvoids prematurely by a mechanism for catastrophic cracking (where these estimates of crack length c are obtained from equation (2.9) with $A \sim 1$).

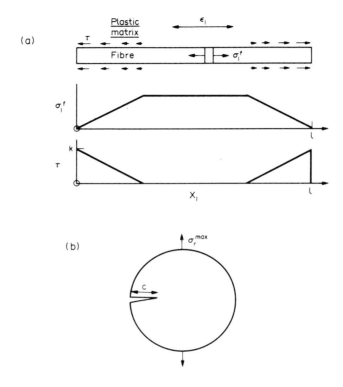

Fig. 2.5. (a) The fibre-loading effect for an elongated hard particle in a
              plastic matrix, leading to the build-up of high tensile stress
              in the central region of the fibre.
          (b) A partially damaged particle containing a crack nucleus of
              length c.

This mechanism of premature particle cracking can therefore operate in any particle of sufficient size to contain critical microcrack lengths from 0.3 to 8 μm and thus
applies to particles with diameters in excess of $\sim$ 5 μm. With this mechanism of
microvoid nucleation an increase in particle size will promote premature cracking,
because the probability of a particle containing a microcrack of critical length will
increase with particle volume.

The mechanisms of microvoid nucleation for the second population N_2 of damaged particles in steels, outlined above, apply in principle to all ductile alloy systems with hard particles. However, the morphology and the mechanical and physical properties of second-phase particles and inclusions in the various alloy systems show such wide variations that it seems unlikely that it will be possible to

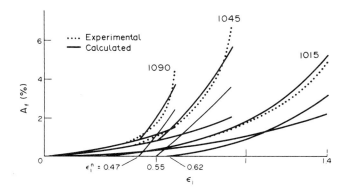

Fig. 2.6.   A comparison of experimental values of microvoid area-fractions A_f, for increasing plastic strain ϵ_1, with theoretical results from a dual-population model. Uniaxial tension tests on spheroidised steels. (After G. Le Roy et al [8]).

develop a general theory of microvoid nucleation. The complexity of each particular alloy/particle system requires an individual quantitative model of microvoid nucleation based on at least two populations representing initially undamaged and initially damaged or elongated particles, respectively. A dual-population model of microvoid nucleation in spheroidised low-alloy steels was recently presented by Le Roy et al. [8], to account for a disproportionate increase in the area fraction of microvoids above a certain critical plastic strain, and a typical set of results is shown in Fig. 2.6.

In addition to the models of microvoid nucleation, discussed above, it is also necessary to develop a model for the apparent non-sequential nature of some void nucleation and growth processes [10]. These effects are concerned with the possibility of a mechanism for spontaneous microvoid nucleation and coalescence (Chapter 1.6) which appears to operate under certain experimental conditions [10]. A model which can account for these effects can be developed from models of microvoid coalescence, and these will be presented together in Chapters 3 and 4.

2.2   The Representation of Microvoid Nucleation Criteria in Principal Stress Space.

In preliminary attempts to develop quantitative models for the nucleation of microvoids at the sites of second-phase particles and inclusions it is helpful to consider the representation of microvoid-nucleation criteria in principal stress space (σ_1, σ_2, σ_3). In the present section we confine attention to microvoid nucleation models for

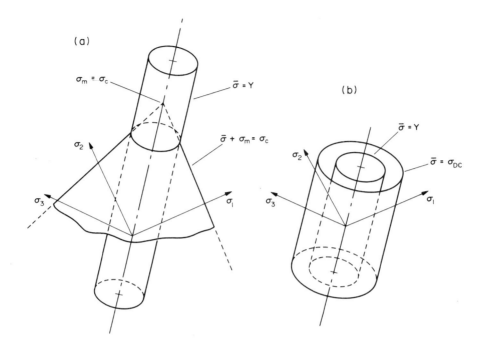

Fig. 2.7.   The initial yield surface in principal stress space $(\sigma_1, \sigma_2, \sigma_3)$
showing (a) the conical surface respresenting the conditions for
microvoid nucleation by matrix/particle decohesion, and (b) the
proposed cylindrical surface representing a critical shear-stress
condition for particle damage at low values of mean-normal
stress σ_m

the N_1 population of strong and firmly-bonded particles which are of approximately equiaxed form and free from internal crack-like defects.

From equations (2.6) and (2.8) the critical condition for particle/matrix decohesion
can be rewritten in the form:

$$(\sigma_1 - \sigma_2)^2 + (\sigma_1 - \sigma_3)^2 + (\sigma_2 - \sigma_3)^2 = 2(\sigma_c - \sigma_m)^2. \qquad (2.10)$$

This equation is represented by a conical surface in principal stress space, Fig.
2.7(a), with the apex situated on the mean-normal stress axis at the point $\sigma_m = \sigma_c$.
The conical surface intersects the initial yield surface $\bar{\sigma} = Y$ at $\sigma_m = \sigma_c - Y$, Fig.
2.7(a), and this implies that whenever the mean-normal stress $\sigma_m \geqslant (\sigma_c - Y)$ decohesion will occur at the very onset of plastic flow. Below the critical level $\sigma_m = (\sigma_c - Y)$,

plastic flow and work-hardening effects are needed to expand the loading surfaces (Chapter 1.4) and elevate the maximum stress at the matrix/particle interface to the value of σ_c for incipient decohesion. It is clear from the form of Fig. 2.7(a) that at low values of mean-normal stress, $\sigma_m \leqslant 0.5Y$, an alternative criterion of microvoid nucleation may become necessary because the expansion of the loading surfaces by work-hardening effects may be insufficient to achieve the critical decohesion stress σ_c. In this case a critical deviatoric or shear stress criterion of particle damage may be achieved first and this could be represented by a cylindrical surface in principal-stress space $(\sigma_1, \sigma_2, \sigma_3)$, Fig. 2.7(b), having an equation of the form:

$$\bar{\sigma} = \sigma_{DC} ; \qquad (2.11)$$

where σ_{DC} is a proposed critical deviatoric stress for particle damage. The critical stress σ_{DC} would need to be estimated for each particle type by experiment and might prove to be either a constant or a function of mean normal stress σ_m. In the latter case, $\bar{\sigma} = \sigma_{DC}(\sigma_m)$, the surface would still be symmetrical about the σ_m axis but would not be cylindrical.

A quantitative illustration of the way in which the critical stress equations (2.10) and (2.11) for microvoid nucleation can be applied in practice is shown in Fig. 2.8, for the case of a uniaxial tension test, with the experimental results of Le Roy et al. [8] for a spheroidised 1045 steel; where $Y = 302$ MPa and $\sigma_c = 1200$ MPa. In uniaxial tension the stresses at the centre of the external neck are $(\sigma_1, \sigma_2 = \sigma_3)$ and equations (2.10) and (2.11) therefore reduce to $\sigma_1 = (3\sigma_c + \sigma_2)/4$ and $\sigma_1 - \sigma_2 = \sigma_{DC}$, respectively; thus when $\sigma_c = 1200$ MPa we obtain $\sigma_1 = (900 + \sigma_2/4)$ MPa. The construction of the microvoid-nucleation loci in Fig. 2.8 is completed on the arbitrary assumption that the critical deviatoric stress for particle damage has a magnitude of the order $\sigma_{DC} \approx 0.75\sigma_c$, giving a coincidence of the two loci for a state of pure tension $(\sigma_1, \sigma_2 = \sigma_3 = 0)$, Fig. 2.8.

In the work of Le Roy et al. [8], where the carbide particles in the spheriodised 1045 steel were sufficiently small ($r \approx 0.5 \mu m$) for the validity of dislocation models of microvoid nucleation [1,2], the critical nucleation-strain equation (2.5) was used to estimate the location of the void-nucleation locus and this was found to be in good agreement with experimental results, Fig. 2.9. The void-nucleation loci shown in Fig. 2.8 for the 'critical stress' continuum models are also replotted on Fig. 2.9 as dotted lines, for comparison. Taking into account the present arbitrary nature of the critical deviatoric-stress region of the void-nucleation locus, it is clear that there is again substantial agreement with the experimental results. This comparison of the 'critical strain' and 'critical stress' criteria of microvoid nucleation is of course only possible because the spheroidised carbides have diameters coinciding with the upper and lower limits of validity, respectively, of the dislocation and continuum models.

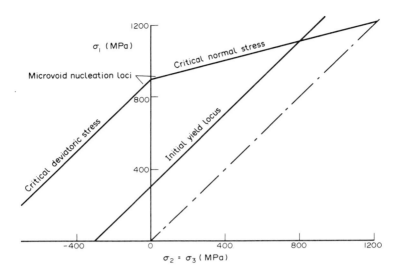

Fig. 2.8. Critical-stress form of the microvoid nucleation loci for a
spheroidised 1045 steel, in a uniaxial tension test, with $\sigma_c = 1200$
MPa and $Y = 302$ MPa; data from G. Le Roy et al [8].

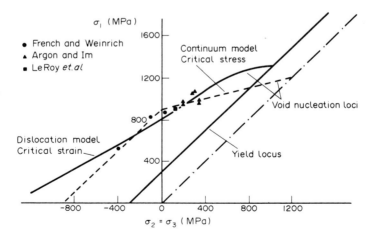

Fig. 2.9. Critical-strain form of the microvoid nucleation locus for a
spheroidised 1045 steel compared with experimental results.
Also showing the critical-stress form of the nucleation loci from
Fig. 2.8. (After G. Le Roy et al [8]).

2.3   The Growth of Microvoids in Plastic-Flow Fields under Arbitrary Plastic-Stress States

Once a microvoid has been nucleated in a plastically deforming matrix, by either the debonding or cracking of a second-phase particle or inclusion, the resulting stress-free surface of the void causes a localised stress and strain concentration in the adjacent plastic field. With continuing plastic flow of the matrix the microvoid will therefore undergo a volumetric growth and shape change which amplifies the distortion imposed by the remote uniform strain-rate field. Under most practical conditions [5,10-12,17] microvoids are nucleated at positions sufficiently far apart for there to be virtually no initial interaction between their local stress and strain-rate fields; it is therefore possible to develop an adequate model for the early stages of microvoid growth in terms of a single void in an infinite plastic solid. The most successful and versatile model of this type was developed by Rice and Tracey [21] for the case of a spherical void of radius R in a remote uniform strain-rate field $\dot{\epsilon}_{ij}$ and remote stress field $\sigma_{ij} = S_{ij} + \sigma_m \delta_{ij}$ (Fig. 2.10), where the strain-rate field is characterised, in terms of the principal components $\dot{\epsilon}_1 \geqslant \dot{\epsilon}_2 \geqslant \dot{\epsilon}_3$, by the Lode variable v [9] defined by:

$$v = -\frac{3\dot{\epsilon}_2}{\dot{\epsilon}_1 - \dot{\epsilon}_3} \quad . \tag{2.12}$$

The analysis [21] was developed in terms of a rigid-plastic non-hardening material and an approximate estimate of the modifying effects of work-hardening was made for the case of an isotropic linear-hardening material. The results for the rates of change $\dot{R}_K$ in the radius of the void, in the directions X_1, X_2, X_3 of the principal strain-rates in the remote strain-rate field, were shown to have the form:

$$\dot{R}_K = \left\{ (1 + E)\dot{\epsilon}_K + (\tfrac{2}{3}\dot{\epsilon}_L\dot{\epsilon}_L)^{1/2} D \right\} R \quad ; \tag{2.13}$$

where $(K, L) = 1, 2, 3$

$(1 + E) \approx 5/3$ for linear hardening, and low values of σ_m with non-hardening,

$(1 + E) \approx 2$ for high values of σ_m with non-hardening,

$D = 0.75 \, \sigma_m/Y$ for linear hardening,

$D = 0.558 \sinh(\frac{3}{2} \frac{\sigma_m}{Y}) + 0.008v \cosh(\frac{3}{2} \frac{\sigma_m}{Y})$ for non-hardening .

The void growth-rate equation (2.13) can be used to give useful estimates of the changes in shape and volume of voids, in various plastic flow fields, by means of an approximate integration procedure. It must be emphasised, however, that the method is only valid for plastic-flow in which the principal axes of the strain-rates

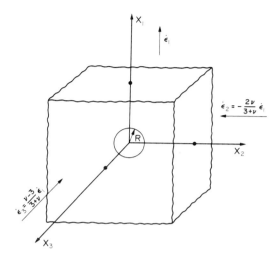

Fig. 2.10. A spherical void in a rigid/plastic solid subjected to remote
stress σ_{ij} and strain-rate $\dot{\epsilon}_{ij}$ fields, showing the principal axes X_i
of the strain rates.

remain fixed in direction throughout the strain path. Nevertheless, this restricted
class of results contain many important flow fields including uniaxial tension ($\nu = +1$), pure shear ($\nu = 0$) and biaxial tension ($\nu = -1$). Before integrating equation
(2.13) it is useful to generalise the results by rewriting the principal strain rates ($\dot{\epsilon}_1$,
$\dot{\epsilon}_2$, $\dot{\epsilon}_3$) entirely in terms of the maximum value $\dot{\epsilon}_1$ and the Lode variable. This is done
with the aid of the remote incompressibility equation $\dot{\epsilon}_1 + \dot{\epsilon}_2 + \dot{\epsilon}_3 = 0$ and equation
(2.12) to give the following expressions for $\dot{\epsilon}_2$ and $\dot{\epsilon}_3$ (Fig. 2.10):

$$\dot{\epsilon}_2 = \frac{-2\nu}{3 + \nu}\ \dot{\epsilon}_1 \ ; \tag{2.14a}$$

$$\dot{\epsilon}_3 = \frac{\nu - 3}{3 + \nu}\ \dot{\epsilon}_1 . \tag{2.14b}$$

Equation (2.13) strictly applies to the case of a perfectly spherical void and in integrating the equation under conditions where the void no longer remains spherical
it is necessary to replace R in equation (2.13) by R_{mean}; where $R_{mean} = 1/3\ (R_1 + R_2 + R_3)$ is the mean value of the principal 'radii' of the resulting ellipsoid. The corresponding rate of change in the mean void-radius is given by $\dot{R}_{mean} = 1/3(\dot{R}_1 + \dot{R}_2 + \dot{R}_3)$ and on combining this with equations (2.13) and (2.14) the resulting equation
can be integrated to give the following expression for the current mean radius of
the void:

$$R_{mean} = \exp\left(-\frac{2\sqrt{3+v^2}}{(3+v)} \, D\epsilon_1\right) R_0 ; \qquad (2.15)$$

where R_0 is the radius of the initial spherical void and ϵ_1 is the total logarithmic plastic strain [9] in the direction of the principal axis X_1. On replacing R by R_{mean} in equation (2.13) the resulting equations can be integrated to give the following general expressions for the three principal radii of the 'ellipsoidal' void:

$$R_1 = \left(A + \frac{(3+v)}{2\sqrt{v^2+3}} \, B\right) R_0 , \qquad (2.16a)$$

$$R_2 = \left(A - \frac{v\,B}{\sqrt{v^2+3}}\right) R_0 , \qquad (2.16b)$$

$$R_3 = \left(A + \frac{(v-3)\,B}{2\sqrt{v^2+3}}\right) R_0 ; \qquad (2.16c)$$

where

$$A = \exp\left(\frac{2\sqrt{v^2+3}}{(3+v)} \, D\epsilon_1\right),$$

$$B = \left(\frac{1+E}{D}\right)(A-1) ,$$

and ϵ_1 is the total logarithmic strain integrated over the total strain path from the initial undeformed state to the current void geometry (R_1, R_2, R_3).

Equations (2.16) have been evaluated for a wide range of mean-normal stresses σ_m/Y, in both the non-hardening and linear-hardening condition, and for Lode variables of $v = +1$ (uniaxial tension) and $v = 0$ (pure shear and plane strain). The mean-normal stress was assumed to be constant over the total strain paths so the results do not apply directly to practical tension tests, where σ_m/Y increases as the external neck develops. Nevertheless, the results for the ratios of the principal dimensions of an ellipsoidal void ($a/a_o = R_1/R_0$, $b/b_o = R_2/R_0$, $c/c_o = R_3/R_0$), which are plotted against the remote maximum principal strain ϵ_1 in Fig. 2.11, give a clear indication of the sensitivity of void growth effects to both the mean-normal stress level and the state of hardening in the matrix.

The results in Fig. 2.11 for constant mean-normal stresses of $\sigma_m/Y = 0.8$ and $\sigma_m/Y = 0.333$, respectively, represent the approximate upper and lower bound curves for void growth in uniaxial tension ($v = +1$), where the mean-normal stress at the centre of an external neck increases with the severity of necking from 0.333 to a value of the order of 0.8 at about 76% reduction in area [22]. The non-hardening results (Fig. 2.11a) show that at low mean-normal stresses the extensional growth of a void a/a_o is greater than that imposed by a purely uniform strain-rate field $\dot{\epsilon}_{ij}$ by a factor of ≈ 1.24 for $\sigma_m/Y = 0.333$ and a factor of ≈ 1.64 for $\sigma_m/Y = 0.8$; with linear hardening (Fig. 2.11b) these amplification factors for a/a_o are reduced to ≈ 1.17 and ≈ 1.47, respectively. The transverse growth of a void ($b/b_o = c/c_o$) shows little amplification over that imposed by the uniform strain-rate field, for mean normal

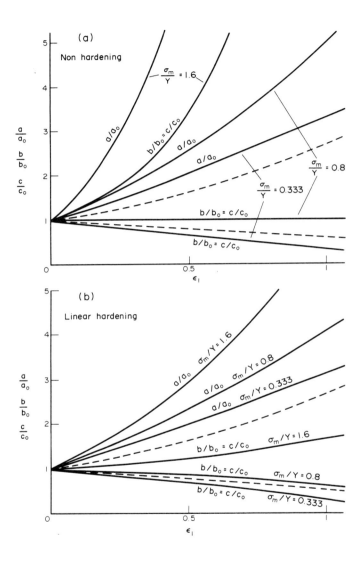

Fig. 2.11. The effect of increasing plastic strain ϵ_1 on the principal radii
(a,b,c) of an initially spherical void ($a_o = b_o = c_o$) in uniaxial tension ($\nu = +1$) at various mean-normal stresses σ_m/Y. (a) Non-
hardening and (b) linear hardening material. The dotted lines
show the changes in void dimensions that would occur without
any amplification of the basic uniform-strain field ($\epsilon_1, \epsilon_2, \epsilon_3$).

stresses up to $\sigma_m/Y = 0.8$, in both the non-hardening and linear-hardening cases, Figs. 2.11(a) and (b); in fact $b/b_o = c/c_o$ is found to either remain constant or *decrease* with increasing total strain. Of course, in real materials any transverse void-closure effect will normally be prevented by the microstructural particles at which the microvoids were nucleated.

The sensitivities of void growth rates to high values of mean-normal stress, in equations (2.16), are illustrated by the results for $\sigma_m/Y = 1.6$ in uniaxial tension ($\nu = +1$), Fig. 2.11, and for $\sigma_m/Y = 2.4$ in plane-strain tension ($\nu = 0$), Fig. 2.12, where the latter σ_m value represents the maximum stress level achieved in the plastic zone ahead of a sharp notch or crack. For uniaxial tension at high mean-normal stress, $\sigma_m/Y = 1.6$, the amplification of extensional growth a/a_o over that imposed by the uniform strain-rate field is given by a factor of ≈ 3.39 with a non-hardening matrix and a factor of ≈ 1.64 with linear hardening; the corresponding amplification factors for transverse growth of the void ($b/b_o = c/c_o$) are ≈ 3.19 for non-hardening and ≈ 1.44 for linear hardening. These results suggest that, at high mean-normal stress levels, matrix work-hardening effects will have a strong influence in reducing the amplification in void growth rates and this effect is likely to be particularly pronounced at very high mean-normal stress $\sigma_m/Y \approx 2.4$, Fig. 2.12, where the non-hardening results exceed the corresponding linear-hardening results in the ratio of $\approx 3:1$.

An assessment of the validity of equations (2.16) as a general model of void growth can be made by comparing the present theoretical results with the experimental results of Atkinson [23,24] for void nucleation and growth in a Cu-SiO$_2$ alloy system. Atkinson observed by T.E.M. the nucleation of voids at SiO$_2$ particles, by debonding at the poles, and the subsequent extensional void growth with increasing plastic strain ϵ_1. When a spherical SiO$_2$ particle separates from the copper matrix, the effective void length in the *matrix* is (2L + D), Fig. 2.13, where D is the particle diameter and 2L is the sum of the lengths of the void nuclei in the direction of the maximum principal strain ϵ_1. The experimental parameter (2L + D)/D is therefore equivalent to the void extensional ratio $a/a_o = R_1/R_o$ from equation (2.16a) and Fig. 2.11. The experimental results of Atkinson are compared with the theoretical results from equations (2.16) in Fig. 2.13, where Atkinson's observation that a nucleation strain $\epsilon_1^n \approx 0.08$ was needed to initiate particle debonding has been taken into account when plotting the theoretical values of a/a_o. The upper and lower bound curves for $\sigma_m/Y = 0.8$ and $\sigma_m/Y = 0.333$, respectively, in both non-hardening and linear-hardening states, show a good general agreement with the experimental results and any slight discrepancies can be accounted for by the fact that the SiO$_2$ particles tended to remain bonded to the matrix in the equatorial regions, thus preventing transverse void growth ($b/b_o = c/c_o$) and tending to restrain the extensional growth (2L + D)/D.

Further confirmation of the general validity of the Rice-Tracey equations for void growth has been obtained by Le Roy et al. [8] who carried out a numerical integration

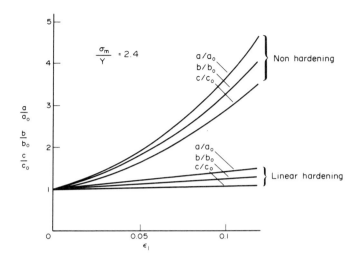

Fig. 2.12. The effect of increasing plastic strain ϵ_1 on the principal radii (a,b,c) of an initially spherical void ($a_o = b_o = c_o$) in plane-strain tension ($\nu = 0$) at the high mean-normal stress $\sigma_m/Y = 2.4$; non-hardening and linear-hardening material.

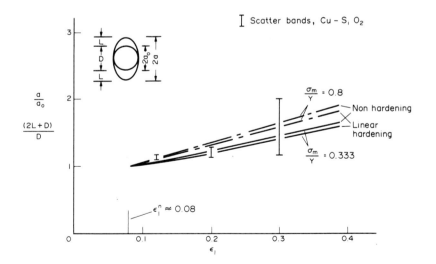

Fig. 2.13. A comparison of the theoretical void-growth ratios a/a_o, in the direction of the maximum principal strain ϵ_1, with the experimental results of Atkinson [23,24] for a Cu-SiO$_2$ Alloy system in which the microvoid nucleation strain $\epsilon_1^n \approx 0.08$.

of the Rice-Tracey equations to give estimates of the changing area-fraction of voids A_f on a microsection, with increasing plastic strain. The theoretical results [8] were obtained for tension tests on materials which were assumed to display a continuous linear increase in microvoid nucleation, and allowances were made for the continuous increase in mean-normal stress as external necking developed in the tension specimens. The experimental results for low-, medium- and high- carbon steels were compared with the corresponding theoretical results, Fig. 2.14, and there was found to be good agreement for the low- and medium-carbon steels up to a strain $\epsilon_1 \sim 0.6$. The authors were able to account for the relatively poor agreement of the results of higher strains by the introduction of a dual-population model of microvoid nucleation [8], Fig. 2.6. It can therefore be concluded that experimental results confirm the general validity of the Rice-Tracey model of void growth [21].

At this stage it is interesting to use the integrated form of the Rice-Tracey model (equations (2.16)) to estimate the apparent departure from constant-volume plastic flow, in a plastic solid containing a certain initial volume fraction V_f° of microvoids. Estimates of an apparent dilational plastic strain ϵ can of course only be obtained on an approximate basis because the remote strain-rate field $\dot{\epsilon}_{ij}$ in the Rice-Tracey model for a single void [21] is incompressible and thus the remote $\epsilon = 0$. However, if we assume that an initially spherical void is situated within a cubic unit cell, Fig. 2.15, of a plastic body and then apply a distant strain-rate field defined by $\dot{\epsilon}_1$ and v, we can obtain a useful estimate of the changing 'dilational' strain ϵ of the *unit cell* with increasing plastic strain ϵ_1 for any chosen value of Lode variable v. From the geometry of the unit cells in Fig. 2.15 it is readily seen that the volumes of the plastic matrix in the initial and the deformed states are given by $(V_o - 4/3 \, \pi \, a_o b_o c_o)$ and $(V_c - 4/3 \, \pi abc)$, respectively, where V_o and V_c are the initial and current volumes of the unit cell. The plastic matrix is incompressible so the two expressions in brackets above are equal, and this results in the following expression for the ratio of the current and initial volumes of the unit cell [25]:

$$V_c / V_o \approx 1 + V_f^{\circ} \left(\frac{abc}{a_o b_o c_o} - 1 \right) ; \qquad (2.17)$$

where the initial volume-fraction of voids $V_f^{\circ} = 4\pi \, a_o b_o c_o / 3 V_o$. The corresponding dilational or mean-normal strain is equal to one-third of the natural-logarithm of the volumetric strain, hence the apparent dilational-plastic strain ϵ is given by:

$$\epsilon \approx 1/3 \ln\left(1 + V_f^{\circ} \left(\frac{abc}{a_o b_o c_o} - 1 \right)\right) ; \qquad (2.18)$$

where a,b,c and a_o,b_o,c_o are the principal radii of the void in the current and initial conditions, respectively.

Equations (2.17) and (2.18) have been evaluated for the case of $V_f^{\circ} = 0.01$, with

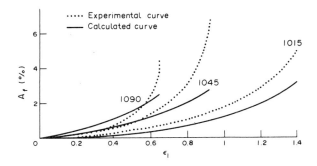

Fig. 2.14. A comparison of the experimental values of microvoid area-
fractions A_f, for increasing plastic strain ϵ_1, with a model for the
continuous nucleation of microvoids [8] based on the Rice-
Tracey model of void growth [21]. Uniaxial tension tests on
spheroidised steels. (After G. Le Roy et al [8]).

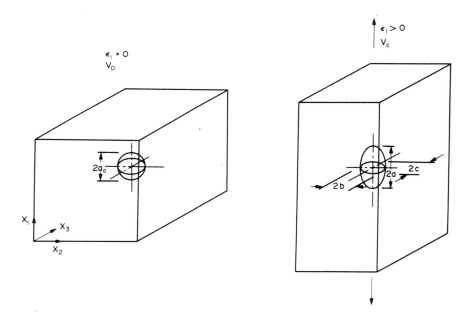

Fig. 2.15. A model for estimating the apparent dilational-plastic strain ϵ in
a body containing microvoids, showing the initial cubic unit-
cell V_0 with a spherical void ($a_0 = b_0 = c_0$) and the deformed
state of the cell V_c at a plastic strain ϵ_1.

various values of σ_m/Y and v, for both non-hardening and linear-hardening conditions; the results for the variation in the dilational plastic strain ϵ with increase in
total plastic strain ϵ_1 are given in Fig. 2.16. These results clearly show that for both
non-hardening and linear-hardening cases, with mean-normal stresses up to σ_m/Y
$= 0.8$, the dilational plastic strain does not exceed a value $\epsilon \approx 0.015$ up to a total
plastic strain $\epsilon_1 = 1.0$. At a higher mean-normal stress of $\sigma_m/Y = 1.6$, for both uniaxial tension ($v = +1$) and plane strain ($v = 0$), the dilational plastic strain $\epsilon \approx 0.015$
is reached at a strain $\epsilon_1 \approx 0.56$ for linear hardening and $\epsilon_1 \approx 0.16$ for non-hardening,
Fig. 2.16. At the very high mean normal stress $\sigma_m/Y = 2.4$, in plane-strain ($v = 0$), a
dilational strain $\epsilon \approx 0.015$ is reached at a strain $\epsilon_1 \approx 0.16$ for linear-hardening and
$\epsilon_1 \approx 0.05$ for non-hardening. It is clear from these results that, with the exception of
the non-hardening results for $\sigma_m/Y = 2.4$, the dilational strain ϵ is always a very
small proportion ($0.015 \sim 0.09$) of the total plastic strain ϵ_1. This result is of fundamental importance in the theory of ductile fracture because it shows that under virtually all practical test conditions the dilational plastic strain ϵ for a solid containing
microvoids is likely to be very small.

It has been proved by Berg [26] that the theory of the plastic potential and the normality rule of plastic flow for incompressible solids (Chapter 1.2), where $f(J_2, J_3) =$
0 is the yield function, applies also to the macroscopic response of a solid containing microvoids, where the yield function has the general form $f(I_1, I_2, I_3) = 0$ (Chapter 1.4). It follows from this result that the gradients of the graphs of ϵ against ϵ_1 in
Fig. 2.16 give an approximate measure of the angular orientation $\emptyset$ of the effective
'dilational plastic' strain-increment vector $d\epsilon_{ij}^{Dil}$ to the deviatoric strain-increment
vector $d\epsilon_{ij}^{Dev}$, Fig. 2.17, where $\cos \emptyset = d\epsilon_{ij}^{Dev}/d\epsilon_{ij}^{Dil}$. The gradient $d\epsilon/d\epsilon_1$ at any particular value of ϵ_1, on the graphs in Fig. 2.16, gives an approximate value of $\tan \emptyset$
(Fig. 2.17), and by the normality rule this gives the angular orientation of the tangent plane to the 'dilational yield' surface with respect to the 'deviatoric' von Mises
yield cylinder.

Estimates of the strain-increment gradients $d\epsilon/d\epsilon_1$ are readily obtained from equations (2.17) and (2.18) with the aid of the following substitution:

$$\frac{abc}{a_o b_o c_o} \approx \left[\frac{R_{mean}}{R_o} \right]^3 = \exp \left[\frac{6\sqrt{3 + v^2}}{(3 + v)} D \epsilon_1 \right] . \tag{2.19}$$

On making this substitution in equation (2.18) and then differentiating we obtain
the following expression for the gradient $\tan \emptyset$ of the effective dilational-yield surface:

$$\tan \emptyset = \frac{d\epsilon}{d\epsilon_1} = \frac{\frac{2\sqrt{3 + v^2}}{(3 + v)} D}{1 - \left[\frac{1 - V_f^o}{V_f^o} \right] / \exp \left[\frac{6\sqrt{3 + v}}{(3 + v)}^2 D \epsilon_1 \right]} , \tag{2.20}$$

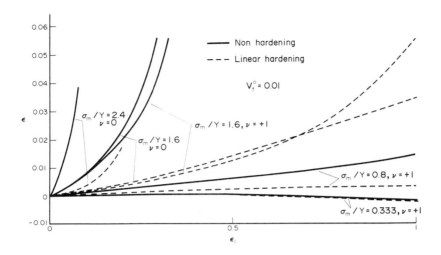

Fig. 2.16. Estimates of the changes in dilational plastic strain ϵ, with increase in the total plastic strain ϵ_1, for an initial volume fraction of voids $V_f^o = 0.01$. Showing the effects of various mean-normal stresses σ_m/Y in both uniaxial tension ($\nu = +1$) and plane-strain tension ($\nu = 0$) for non-hardening and linear-hardening materials.

where D is defined in equation (2.13). This expression for tan ϕ was used to obtain the gradients of the dilational-yield surface at various mean-normal stress levels σ_m/Y for the case of uniaxial tension ($\nu = +1$) and an initial volume fraction of voids $V_f^o = 0.01$. The results were obtained for both non-hardening and linear-hardening conditions, at strains of $\epsilon_1 = 0$ and $\epsilon_1 = 0.2$, and are given in Table 2.1.

It is therefore possible to obtain an estimate of the departure of the effective 'dilational' yield surface, from the von Mises yield surface for the intervoid matrix, by fitting a polynomial expression to the gradients in Table 2.1 of the form:

$$\frac{d(\bar{\sigma}/Y)}{d(\sigma_m/Y)} = a_1 + 2a_2 \left(\frac{\sigma_m}{Y}\right) + 3a_3 \left(\frac{\sigma_m}{Y}\right)^2 + 4a_4 \left(\frac{\sigma_m}{Y}\right)^3 + 5a_5 \left(\frac{\sigma_m}{Y}\right)^4 . \tag{2.21}$$

After evaluating the constants $a_1 \dots a_5$ the equation is integrated to give an expression for the radius of the 'dilational' yield surface as a function of σ_m/Y:

$$\bar{\sigma}/Y = a_o + a_1 \left(\frac{\sigma_m}{Y}\right) + a_2 \left(\frac{\sigma_m}{Y}\right)^2 + a_3 \left(\frac{\sigma_m}{Y}\right)^3 + a_4 \left(\frac{\sigma_m}{Y}\right)^4 + a_5 \left(\frac{\sigma_m}{Y}\right)^5 . \tag{2.22}$$

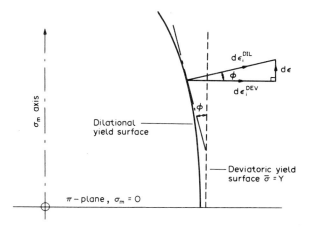

Fig. 2.17. Angular orientation ϕ of the effective dilational-plastic yield surface and associated strain-increment vector $d\epsilon_i^{Dil}$ to the corresponding deviatoric (von Mises) yield surface and strain-increment vector $d\epsilon_i^{Dev}$. Principal-stress space (σ_1, σ_2, σ_3) representation with the mean-normal stress σ_m and dilational strain-increment $d\epsilon$ axes in a vertical direction.

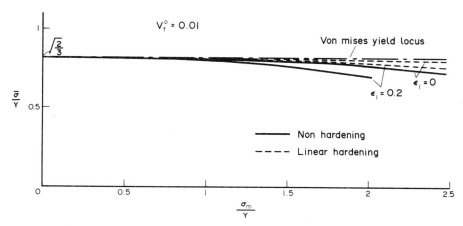

Fig. 2.18. Effective dilational-plastic yield loci for an initial volume fraction of voids $V_f^o = 0.01$, at tensile plastic strains of $\epsilon_1 = 0$ and $\epsilon_1 = 0.2$ in both non-hardening and linear-hardening materials.

Table 2.1   Values of the gradient tan ø to the effective dilational-yield surface for an initial volume fraction of voids $V_f^\circ = 0.01$. Uniaxial tension ($\nu = +1$) for incipient plasticity ($\epsilon_1 = 0$) and large plastic flow ($\epsilon_1 = 0.2$) at various mean-normal stresses σ_m .

	σ_m/Y	tan ø (non-hardening)	tan ø (linear-hardening)
$\epsilon_1 = 0$	0	~0.00008	0
	0.333	0.003	0.0025
	0.8	0.0086	0.006
	1.6	0.031	0.012
	2.4	0.1035	0.018
$\epsilon_1 = 0.2$	0	~0.00008	0
	0.333	0.0036	0.0029
	0.8	0.0142	0.0086
	1.6	0.1880	0.0244
	2.4		0.0520

The value of a_0 is obtained by assuming that the radius of the 'dilational' yield surface at $\sigma_m/Y = 0$ is related to that of the von Mises surface by a simple "law of mixtures" expression. The effective dilational-yield surface obtained in this way, from the results in Table 2.1, are used to plot the yield loci in Fig. 2.18. These results show that for mean-normal stress levels up to $\sigma_m/Y \approx 1.2$ there is likely to be very little departure from a deviatoric plastic response for $V_f^\circ \geq 0.01$. As the mean-normal stress level is raised to $\sigma_m/Y \approx 2.4$ there is a significant increase in dilational plasticity for the non-hardening solid but with linear-hardening the response remains essentially deviatoric. The results in Fig. 2.18 will be discussed in more detail in Chapter 3 in relation to dilational-plastic models of ductile fracture.

REFERENCES

1. Brown, L.M. and Stobbs, W.M. *Phil. Mag*, 1976, 34, p.351.
2. Goods, S.H. and Brown, L.M. *Acta Metall*, 1979, 27, p.1.
3. Argon, A.S., Im, J. and Safoglu, R. *Metallurgical Trans,* 1975, 6A, p.825.
4. Argon, A.S. and Im, J. *Metallurgical Trans.,* 1975, 6A, p.839.
5. Van Stone, R.H., Cox, T.B., Low, J.R. Jr. and Prioda, J.A. *Int. Metals Reviews,* 1983, 30, p.157.
6. Gurland, J. and Plateau, J. *Trans. ASM,* 1963, 56, p.442.
7. Tanaka, K., Mori, T. and Nakamura, T. *Phil. Mag.,* 1970, 21, p.267.
8. Le Roy, G., Embury, J.D., Edwards, G. and Ashby, M.F. *Acta Metall,* 1981, 29, p.1509.
9. Hill, R. *The Mathematical Theory of Plasticity,* Clarendon Press, Oxford, 1950.
10. Wilsdorf, H.G.F. *Mat. Sci. Eng.,* 1983, 59, p.1
11. Leslie, W.C. *I.S.S. Trans.* 1983, 2, p.1.
12. Cox, T.B. and Low, J.R. Jr. *Metallurgical Trans.,* 1974, 5, p.1457.
13. Cox, H.L.   Brit. *J. Appl. Physics,* 1952, 3, p.72.
14. Barnby, J.T. *Acta Metall,* 1967, 15, p.903.
15. McMahon, C.J. and Cohen, M. *Acta Metall,* 1965, 13, p.591.
16. Cox, T.B.   PhD Thesis, Carnegie-Mellon Univ., Pittsburgh, PA, 1973.
17. Gurland, J. *Acta Metall,* 1972, 20, p.735.
18. Liu, C.T. and Gurland, J. *Trans. ASM,* 1968, 61, p.156.
19. Kelly, A. and Macmillan, N.H. *Strong Solids,* Clarendon Press, Oxford, 1986.
20. Webb, W.W. and Forgery, W.D. *Acta Metall,* 1958, 6, p.462.
21. Rice, J.R. and Tracey, D.M. *J. Mech. Phys. Solids,* 1969, 17, p.201.
22. Argon, A.S., Im, J. and Needleman, A. *Metallurgical Trans.,* 1975, 6A, p.815.
23. Atkinson, J.D.   PhD Thesis, University of Cambridge, 1973.
24. Brown, L.M. *Cambridge Fracture Conf.* Churchill College, Institute of Physics, 1974.
25. Thomason, P.F. *Acta Metall,* 1985, 33, p.1087.
26. Berg, C.A. *Inelastic Behaviour of Solids*, (edited by M.F. Kanninen et al.), McGraw-Hill, 1970, p.171.

CHAPTER 3

The Mechanics of Ductile Fracture by the Nucleation, Growth and Coalescence of Microvoids

3.1 Ductile Plastic Flow in Metals Containing Microvoids

It was emphasised in Chapter 2 that engineering alloy systems generally contain inclusions and second-phase particles of widely varying morphologies and mechanical properties. Consequently it is not strictly possible to define a unique microvoid-nucleation strain ϵ_1^n in any metal undergoing a continuing process of plastic flow. The microvoid nucleation strain ϵ_1^n for the individual particles types(N_1, N_2, ...) in a particular alloy system, will usually be achieved at different points in a continuing plastic-flow process, and once a set of microvoids has been nucleated, at the sites of a particular type of particle, the microvoids will immediately begin the continuous extensional and volumetric growth process induced by the remote applied stress and strain-rate fields (cf. Chapter 2.3). The microvoid growth processes for the various particle types (N_1, N_2, ...) will continue until a certain critical growth strain ϵ_1^G is reached for the dominant population N_i, whereupon microvoid coalescence will occur across a sheet of microvoids by localised internal microscopic necking (plastic limit-load failure) of the intervoid matrix (cf. Chapter 1.6). The effective plastic strain-increment at the macroscopic level, associated with the complete internal microscopic-necking process at a sheet of microvoids, will normally be of negligible size in comparison to the total growth strain ϵ_1^G. It follows, therefore, that the macroscopic ductile-fracture strain ϵ_1^F in a metal containing microstructural particles can be represented by:

$$\epsilon_1^F\big|_{N_i} = \epsilon_1^n\big|_{N_i} + \epsilon_1^G\big|_{N_i} \;; \tag{3.1}$$

where N_i is the particle population which has the controlling influence on ductile fracture. That is, the population N_i that requires the least strain to initiate ductile fracture by internal microscopic necking. It must of course be remembered when applying equation (3.1) that there is a possibility of an interaction between the particle populations N_i, in some situations, in which case the fracture strain would not have either of the simple forms $\epsilon_1^F|_{N_1}$ or $\epsilon_1^F|_{N_2}$. The ductile fracture strain with population-interaction effects $\epsilon_1^F|_{N_1,N_2}$ could only be evaluated by a detailed analysis of the effects of each particle type on the nucleation, growth and coalescence of voids in a particular alloy/particle system; the predominant interaction between the microvoid populations occurring at the point of incipient void coalescence.

It is clear from the form of the ductile-fracture strain equation (3.1) that in addition to the theories of microvoid nucleation and growth, presented in Chapter 2, there is a need for a criterion of microvoid coalescence to establish the *magnitude* of the void-growth strain $\epsilon_1^G|_{N_i}$ which will produce a state of incipient ductile fracture in a particular microvoid population N_i, and thus give a method of estimating the total plastic strain to ductile fracture $\epsilon_1^F|_{N_i} = \epsilon_1^n|_{N_i} + \epsilon_1^G|_{N_i}$. The present chapter is therefore primarily concerned with the problem of establishing the critical conditions which will lead to the intervention of ductile fracture, by a localised mode of internal microscopic necking across a sheet of microvoids, in a previously homogeneous state of plastic flow for a solid containing a single void population N_i. We therefore implicitly neglect any possible interactions between the microvoid populations N_1 and N_2, and assume that either N_1 or N_2 separately takes on the controlling influence on the ductile-fracture strain ϵ_1^F. The effects of possible microvoid-population interactions will be considered separately at a later stage.

At the macroscopic level the sudden intervention of a ductile-fracture surface, in a previously homogeneous plastic-flow field, is mathematically equivalent to the development of a stationary velocity-discontinuity in the plastic velocity-field. It is therefore relevant at this stage to consider the work of R. Hill [1], on discontinuity relations in solid mechanics, which gives an invariant formulation of the conditions for establishing the location of the characteristic surfaces, in a plastic velocity field, on which stationary velocity-discontinuities or fracture surfaces can develop.

3.2 Characteristic Surfaces in Rigid/Plastic Velocity Fields and their Relation to the Mechanics of Ductile Fracture.

It has been shown by Hill [1] that for a continuous velocity field u_i in a deformable solid, it may be possible to find characteristic surfaces Σ along which u_i is constant and certain first derivatives, with components normal to Σ, remain undetermined. Hill obtained the condition for the existence of a velocity-field characteristic in terms of tensor strain-rates $\dot{\epsilon}_{ij}$ on a characteristic segment decomposed into normal and tangential components, and this was found to have the form:

$$\dot{\epsilon}_{ij} = \dot{\epsilon}_N \nu_i \nu_j + \dot{\gamma}_T(\mu_i \nu_j + \mu_j \nu_i), \quad \mu_k \nu_k = 0; \qquad (3.2)$$

where v_i and μ_i are unit vectors lying normal and tangential to the surface Σ, respectively, $\dot{\epsilon}_N$ is a pure extension-rate normal to Σ and $\dot{\gamma}_T$ is a tensor shear-rate tangential to Σ. Equation (3.2) is a statement that the tensor strain-rate components lying *in* the characteristic surface Σ are zero and the rest are undetermined. This result can be stated more precisely by choosing rectangular coordinates (xyz) with (x,z) in the plane of Σ and y lying normal to Σ. We then find from condition (3.2) that $\dot{\epsilon}_y$, $\dot{\epsilon}_{yx}$, $\dot{\epsilon}_{yz}$ are undetermined and:

$$\dot{\epsilon}_x = \dot{\epsilon}_z = \dot{\epsilon}_{xz} = 0 . \tag{3.3}$$

This result is of great importance in the theory of ductile fracture since it provides a simple invariant method of locating the characteristic surfaces Σ in a plastic velocity field u_i, where the normal components of the first derivatives of u_i are undetermined by the boundary conditions. The characteristics Σ are therefore the surfaces on which stationary velocity discontinuities, with both normal and tangential components, *can* develop in a previously continuous velocity field; a ductile-fracture surface being mathematically equivalent to a stationary velocity-discontinuity with both normal and tangential components.

It is important to note that Hill derived the equations for the velocity-field characteristics (equations (3.2) and (3.3)) without particular reference to the constitutive equations for the solid. It follows, therefore, that the characteristic condition (3.3) applies both to deviatoric-plastic materials satisfying $f(J_2, J_3) = 0$ (cf. Chapter 1.1) and dilational-plastic materials with $f(I_1, I_2, I_3) = 0$; in addition, the yield surfaces could also contain singularities (or vertexes) where there is no unique normal [1] (cf. Chapter 1.2).

It is instructive, at this point, to apply the characteristic condition (3.3) to a number of simple plastic velocity fields, using the Mohr's strain-rate circle to locate the characteristic surfaces Σ. For the case of a rigid/plastic solid the Prandtl-Reuss flow rule (equation (1.22)) is replaced by the Levy-Mises flow rule [2] and in terms of principal strain-rates $\dot{\epsilon}_i$ this becomes:

$$\dot{\epsilon}_i = \dot{\lambda}S_i = \dot{\lambda}[\sigma_i - \sigma_m] \quad ;$$

where S_i, σ_i, σ_m are defined in Chapter 1.2 and $\dot{\lambda}$ is a scalar rate of proportionality. On applying this equation to the case of a uniaxial tensile stress field ($\sigma_2 = \sigma_3 = 0$) we find that $\dot{\epsilon}_1 = \frac{2}{3}\dot{\lambda}\sigma_1$, $\dot{\epsilon}_2 = \dot{\epsilon}_3 = -\frac{1}{3}\dot{\lambda}\sigma_1$. The principal strain-rates ($\dot{\epsilon}_1$, $\dot{\epsilon}_2 = \dot{\epsilon}_3$) of the plastic strain-rate field now define a Mohr's strain-rate circle (Fig.3.1) and the angular orientation ψ of the characteristic surfaces in the physical plane are defined from condition (3.3) by $\dot{\epsilon}_x = 0$. From the construction of the Mohr's circle the characteristics Σ in both the (1,2) and (1,3) planes are found to have an orientation $\psi = \pm 54.74°$ with respect to the direction of maximum principal strain-rate $\dot{\epsilon}_1$ (Fig. 3.1). In this case, the slip-lines, which are defined as the orthogonal networks coinciding at all points with the directions of maximum shear-strain rate (i.e., at $\pm 45°$ to $\dot{\epsilon}_1$), do

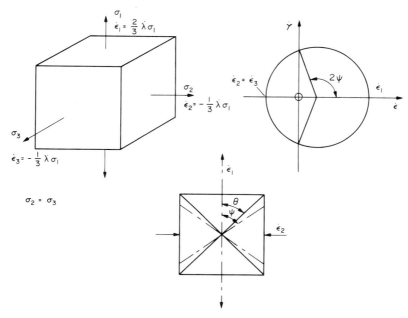

Fig. 3.1. The uniaxial plastic stress and strain-rate fields (a), with the associated Mohr's strain-rate circle (b), and the corresponding orientations (c) of both the characteristic surfaces Σ ($\psi = \pm 54,74°$) and the slip-lines ($\theta = \pm 45°$).

not coincide with the characteristics Σ.

We now consider the plane-strain plastic field ($\dot{\epsilon}_3 = 0$) where $\sigma_3 = \frac{1}{2}(\sigma_1 + \sigma_2)$ and from the Levy-Mises equation $\dot{\epsilon}_i = \dot{\lambda} S_i$ we obtain $\dot{\epsilon}_1 = \frac{1}{2} \dot{\lambda}(\sigma_1 - \sigma_2)$, $\dot{\epsilon}_2 = -\frac{1}{2} \dot{\lambda}(\sigma_1 - \sigma_2)$; thus giving $\dot{\epsilon}_1 = -\dot{\epsilon}_2$. The Mohr's circle construction in Fig. 3.2 shows that for plane-strain plastic velocity-fields the characteristics have the orientation $\psi = \pm 45°$ and thus coincide with the slip-lines. There will therefore always be a tendency for duc-tile-fracture surfaces to form along directions which coincide with the slip-lines in states of plane-strain plastic flow; i.e., at $\pm 45°$ to the direction of the maximum principal strain-rate $\dot{\epsilon}_1$. It must be emphasised however that, although the slip-lines are also directions of maximum shear-stress in isotropic materials, the ductile-frac-ture surfaces which develop along the slip-lines in continuous plane-strain plastic fields are *not* "shear" fractures in the sense that they have developed by a predo-minantly tangential velocity-discontinuity along the characteristic surface. In fact, an incipient "shear" fracture of that type would normally be geometrically incom-patible with the surrounding continuous plane-strain plastic field. Hence, any duc-tile-fracture surface developing along the axis of symmetry of a slip-line field must

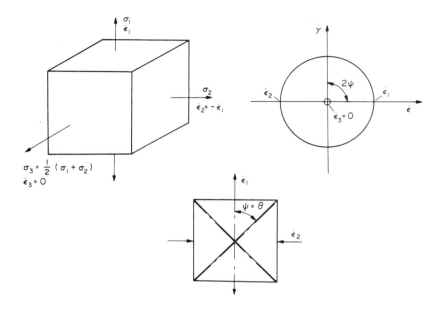

Fig. 3.2. The plane-strain plastic stress and strain rate fields (a), with the associated Mohr strain-rate circle (b), showing the corresponding coincidence (c) of the characteristics Σ and the slip-lines ($\psi = \theta = \pm 45°$).

have normal and tangential components of velocity-discontinuity with equal magnitude; thus giving a resultant velocity-discontinuity in the direction of the existing maximum principal strain-rate to provide geometrical compatibility at incipient fracture. The compatibility relations for the normal and tangential components of velocity-discontinuity at ductile-fracture surfaces, lying along characteristics of arbitrary orientation, are given in the following section. In view of the above comments, it is unfortunate to note that the literature on ductile fracture contains many references to "shear" fractures which are in fact the normal form of microvoid-coalescence surface, with approximately equal normal and tangential components of velocity-discontinuity, lying parallel to the slip-lines (cf. Chapter 1.6). The term "shear fracture" should be reserved for those special types of ductile-fracture surface, which can only develop in plastic fields with relatively low geometrical constraint, where the incompressible-plastic velocity field contains tangential velocity-discontinuities on certain sets of slip-lines. Examples of plastic velocity fields with tangential velocity-discontinuities, where true "shear" fractures can occur, are presented in Chapters 6.5, 7.3 and 8.2.

Although the characteristic surfaces Σ in a plastic velocity-field represent the surfaces on which ductile fracture will tend to develop, they exist at all stages of a plastic-flow process and do not give any information on the critical conditions which lead to incipient ductile fracture. To establish the critical conditions for ductile fracture-surface formation it is necessary to carry out a detailed analysis of the physics of ductile fracture and then develop a mechanical model that represents the physical process of ductile fracture as closely as possible.

3.3 The Mechanics of Ductile Fracture by Microvoid Coalescence

The physical analysis of the ductile-fracture process in a uniaxial tension specimen, presented in Chapter 1.6, shows that ductile-fracture surfaces are formed by the sudden catastrophic coalescence of microvoids which nucleate and grow at the sites of inclusions and second-phase particles. A schematic representation of the complete ductile-fracture process is shown in Fig. 3.3 for (a) the initial undeformed state $\epsilon_1 = 0$, (b) the microvoid nucleation strain $\epsilon_1 = \epsilon_1^n$, (c) the microvoid growth strain $\epsilon_1^n < \epsilon_1 < \epsilon_1^F$, and the three stages of fracture-surface formation at $\epsilon_1 = \epsilon_1^F$ involving (d) incipient microvoid coalescence, (e) internal microscopic necking and (f) knife-edge separation. An important point to note from both Fig. 3.3 and the physical analysis in Chapter 1.6 is that in regions immediately adjacent to a ductile fracture surface the microvoids are still relatively small and widely separated; the *large* transverse growth which brings about complete microvoid coalescence begins only at incipient ductile fracture and is caused almost entirely by localised internal necking of the intervoid matrix at the 'weakest' sheet of microvoids. A further point to note is that the formation of a fracture surface along a characteristic Σ (Fig. 3.4), by microvoid coalescence, introduces a 'jump' or discontinuity Δ in the velocity gradients across the fracture surface, which must have normal Δ_N and tangential Δ_T components satisfying the following compatibility condition:

$$\Delta = \Delta_T \cos^{-1}\psi = \Delta_N \sin^{-1}\psi \quad , \tag{3.4}$$

where ψ is the angular orientation of the characteristic Σ with respect to the direction of maximum principal strain-rate $\dot{\epsilon}_1$. The form of this compatibility equation confirms the point made above that fracture surfaces lying along a characteristic Σ are not to be regarded as 'shear' fractures.

Before developing a detailed model of ductile fracture based on the above observations of the physics of ductile fracture, we first consider the essential features of the mechanics of microvoid coalescence with the aid of a relatively simple two-dimensional plane-strain model (Fig. 3.3) for a state of incipient microvoid coalescence in a work-hardening plastic/rigid solid with a current yield-shear stress k_n for the intervoid matrix material. The sudden transformation from a macroscopically homogeneous state of plastic flow, under the applied stress $\sigma_1 = 2k + \sigma_2$ (where k

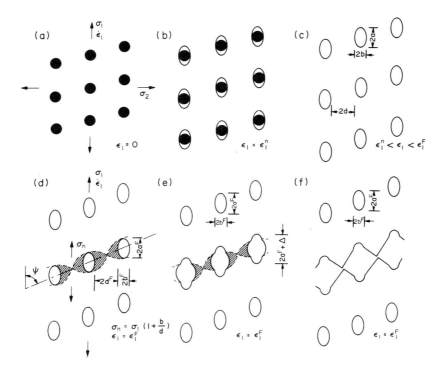

Fig. 3.3. A schematic representation of the ductile-fracture process in metals from the undeformed state (a), through microvoid initiation (b) and growth (c), to the three principal stages of fracture-surface formation: (d) incipient microvoid coalescence; (e) internal microscopic necking and (f) knife-edge separation.

is the current yield-shear stress at the macroscopic level), to the highly localised internal necking of the intervoid matrix across a single 'weakest' sheet of microvoids, coincides [3] with the attainment of the plastic limit-load condition for localised plastic failure of the intervoid matrix. The incipient limit-load condition can be represented in terms of the critical value of the mean stress σ_n which is required to initiate localised plastic flow or internal necking in the intervoid matrix of a 'porous' solid, Fig. 3.3(a). When the microvoids are small and widely spaced, i.e., $a/d \ll 1$ (Fig. 3.3(c)), σ_n is considerably greater than σ_1 and plastic limit-load failure is prevented [3] because:

$$\sigma_n \left(\frac{d}{d+b} \right) > \sigma_1 = 2k + \sigma_2 \ , \tag{3.5}$$

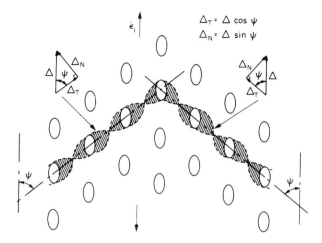

Fig. 3.4. The formation of a ductile-fracture surface, by internal microscopic necking along the characteristic Σ orientations ψ, showing
the 'jump' or discontinuity Δ in the velocity gradients across the
incipient fracture surface and the compatible normal Δ_N and
tangential Δ_T components.

where 2a and 2b are the length and breadth of the microvoids, respectively, and 2d
is the intervoid spacing. With continuing tensile plastic flow in the σ_1 direction the
parameter a/d, representing the geometry of the intervoid-matrix 'neck', gradually
increases (Fig. 3.3(d)) and this has the effect of reducing the mean stress σ_n for plastic limit-load failure of the intervoid matrix (Fig. 3.5) to the point where the current
applied stress σ_1 is sufficient to initiate this mode of localised internal necking. The
critical condition for incipient microvoid coalescence, by plastic limit-load failure of
the intervoid matrix, is therefore given by:

$$\sigma_n(1 - \sqrt{V_f}) = \sigma_1 \quad , \qquad (3.6)$$

where V_f is the volume fraction of voids in the two-dimensional model and $(1 - \sqrt{V_f})$
$\approx$ d/(d + b) [3]. The effects of increasing plastic strain ϵ_1 on both σ_n and σ_1 are shown
schematically in Fig. 3.6, where the point of intersection of the two curves repre-
sents the ductile-fracture condition (3.6) and gives the ductile-fracture strain $\epsilon_1 = \epsilon_1^F$.

It should be emphasised at this point that the plastic strain ϵ_1 at incipient limit-load
failure of the intervoid matrix is virtually equal to the ductile-fracture strain ϵ_1^F
because, once the internal necking mode of microvoid coalescence begins, virtually

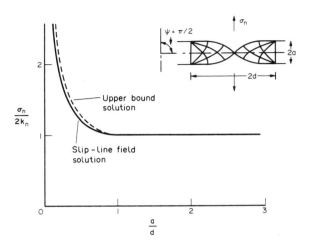

Fig. 3.5. The constraint factors $\sigma_n/2k_n$ for incipient plastic limit-load failure of the intervoid matrix, with variations in the neck-geometry parameter a/d. (After P.F. Thomason, J. Inst. Metals, 1968, 96, p.360).

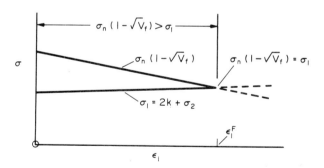

Fig. 3.6. Schematic representation of the effects of increasing plastic strain ϵ_1 on the virtual plastic-limit-load stress $\sigma_n(1 - \sqrt{V_f})$ and the actual stress σ_1; showing the attainment of the critical condition for ductile fracture $\sigma_n(1 - \sqrt{V_f}) = \sigma_1$ at $\epsilon_1 = \epsilon_1^F$.

all the subsequent plastic deformation is confined to the band of velocity discontinuity Δ across the incipient fracture surface (Fig. 3.3 and 3.4), and this is equivalent to an *infinitesimal* plastic-strain increment at the macroscopic level [3]. A further important point to note is that condition (3.6) applies at an instant where the state of hardening is still approximately homogeneous, from the intervoid-matrix level (k_n) to the macroscopic level (k), because the intense localised deformation in the intervoid matrix associated with internal necking is only just about to begin. It follows from this that, for sufficiently small volume fractions V_f of voids, the yield shear stress both in the intervoid matrix and at the macroscopic level can be regarded as approximately equal, i.e., $k_n \approx k$. We can therefore normalise condition (3.6) by the current yield shear stresses $k_n \approx k$ of the work-hardening matrix to give [3].

$$\frac{\sigma_n}{2k}(1 - \sqrt{V_f}) = \frac{\sigma_1}{2k} = \frac{1}{2} + \frac{\sigma_m}{2k}, \tag{3.7}$$

where σ_m is the mean normal stress in the plane-strain model. A criterion for microvoid coalescence similar to equations (3.6) and (3.7) has also been presented by Brown et al. [4] but this neglects the important effects of stress triaxiality (represented by the presence of σ_m in equations (3.6) and (3.7)) and assumes that plastic limit-load failure will only begin when the neck geometry parameter $a/d \sim 1$ and thus $\sigma_n \approx 2k$. Condition (3.7) shows that with high stress triaxiality (i.e., $\sigma_m \gg k$) the condition for incipient microvoid coalescence can be reached with $\sigma_n \gg 2k$, and a/d $\ll 1$.

The assumption made above, in connection with the ductile-fracture criterion (3.7), that the yield shear stress k_n in the intervoid matrix is virtually equal to the 'macroscopic' yield shear stress k is strictly only valid for very low volume fractions V_f of microvoids (i.e., $V_f \sim 0.01$). For volume fractions $V_f > 0.01$, k_n and k can be related approximately by a simple 'law of mixtures' expression of the form:

$$k = (1 - V_f) k_n . \tag{3.8}$$

Combining equations (3.7) and (3.8) we now obtain the following ductile-fracture criterion which is valid for relatively large volume fractions of microvoids (i.e., $0 \leqslant V_f \leqslant 0.1$):

$$\frac{\sigma_n}{2k_n}(1 + \sqrt{V_f})^{-1} = \frac{\sigma_1}{2k} = \frac{1}{2} + \frac{\sigma_m}{2k}. \tag{3.9}$$

It should be emphasised at this point that a simple 'law of mixtures' relation (equation (3.8)) between the macroscopic and microscopic values of the yield shear stress will only be valid when the microvoids are nucleated at the larger non-coherent type of particles in a ductile matrix. For the case of substantial microvoid nucleation at the very small coherent precipitate particles, the reduction in yield stress 'felt' at the macroscopic level would be substantially more than a simple 'law of mixtures' effect. However, as pointed out in Chapter 2.1, in most practical situations microvoid nucleation and growth occurs predominantly at the sites of the larger non-coherent particles, and under these conditions equations (3.8) and (3.9) will be valid.

When developing the model for incipient ductile fracture by microvoid coalescence, represented by the critical condition (3.9), it was proposed that coalescence would always tend to occur on a characteristic surface ψ of the macroscopic velocity field. It is therefore necessary to examine the effects of variations in the orientation ψ of the void-coalescence plane, on the plastic constraint factor $\sigma_n/2k_n$ at incipient limit-load failure of the intervoid matrix, to establish whether the critical conditions are in fact always likely to be achieved first along a characteristic surface. It is very difficult to obtain valid slip-line field solutions for incipient microvoid coalescence on surfaces of arbitrary orientation in the range $0 < \psi \leqslant 90°$, however, a useful set of upper-bound and lower-bound solutions is presented below.

A kinematically admissible velocity field for incipient limit-load failure of the intervoid matrix, on a surface of arbitrary orientation ψ, is shown in Fig. 3.7 and this is used to obtain the rate of internal energy dissipation for the upper-bound solution [2,5]:

$$\dot{I} = \sqrt{2}\, k_n \int_V [\, \dot{\epsilon}_x^2 + \dot{\epsilon}_y^2 + \tfrac{1}{2}\, \dot{\gamma}_{xy}^2 \,]^{\tfrac{1}{2}} \; dV + k_n \int_S \dot{s}\; dS\,, \tag{3.10}$$

where k_n is the yield shear stress of the intervoid matrix, V is the volume of the plastic zone, S is the surface area of the rigid-plastic boundary, and the strain-rate components are related to the velocities by $\dot{\epsilon}_x = \partial \dot{u}/\partial x$, $\dot{\epsilon}_y = \partial \dot{v}/\partial y$, $\dot{\gamma}_{xy} = (\partial \dot{u}/\partial y + \partial \dot{v}/\partial x)$. The expression for $\dot{I}$ is now equated to the rate of work of the external loads $\dot{E} = 4\sigma_n \dot{w} d \sin \psi$ (where the parameters are defined in Fig. 3.7) to give an upper-bound expression for the plastic constraint factor $\sigma_n/2k_n$ as a function of the orientation ψ of the void-coalescence surface. The upper-bound numerical results for the variation of $\sigma_n/2k_n$ with ψ are shown in Fig. 3.8 for microvoid volume fractions of $V_f = 0.02$ and $V_f = 0.05$, respectively. The corresponding lower-bound solutions for $\sigma_n/2k_n$, over the range $\sim\pi/4 \leqslant \psi \leqslant \pi/2$, were obtained by the statically-admissible extension of simple slip-line fields in the intervoid matrix, by the techniques of J.M. Alexander [6]; the results are plotted along with the upper-bound results in Fig. 3.8.

The results in Fig. 3.8 suggest that, within the limits of accuracy of the upper- and lower-bound solutions, the critical constraint factor $\sigma_n/2k_n$ for incipient limit-load failure of the intervoid matrix is likely to be only weakly dependent on the orientation ψ of the incipient fracture surface in the range $\pi/4 \leqslant \psi \leqslant \pi/2$. It follows therefore that ductile fracture surfaces are likely to develop predominantly on the characteristic

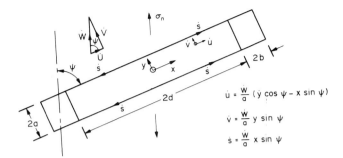

$$\dot{u} = \frac{w}{a} (\dot{y} \cos \psi - x \sin \psi)$$

$$\dot{v} = \frac{w}{a} y \sin \psi$$

$$\dot{s} = \frac{w}{a} x \sin \psi$$

Fig. 3.7. The kinematically admissible velocity field $(\dot{u}, \dot{v})$ for the upper-bound solution to the problem of incipient limit-load failure of the intervoid matrix, along a characteristic surface of orientation ψ; where $\dot{s}$ is a velocity discontinuity.

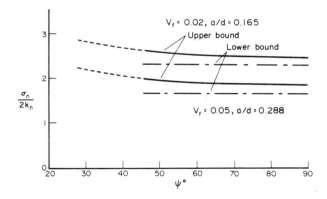

Fig. 3.8. The upper-bound constraint factors $\sigma_n/2k_n$ for incipient limit-load failure on characteristic orientations ψ, with microvoid volume fractions of $V_f = 0.02$ and 0.05, respectively. Also showing the slip-line field solutions at $\psi = \pi/2$ as lower-bounds over the range $\sim\pi/4 \leqslant \psi < \pi/2$.

surfaces Σ of the velocity field (cf. Section 3.2). However, it must be remembered that, although microvoid coalescence by limit-load failure of the intervoid matrix will always *tend* to develop more readily on the characteristic surfaces of a particular plastic field, where compatible discontinuities in the normal and tangential velocities can readily develop, ductile-fracture surfaces are not necessarily *confined* to the characteristics and could under certain circumstances develop on surfaces that are not characteristic. This follows from the fact that the internal-necking mode of microvoid coalescence differs to first order from the current stable mode of plastic flow; this was pointed out previously [7] in connection with some experimental results [8,9] which confirm that ductile-fracture surfaces do not always coincide with the characteristics. It follows, therefore, that although a *sufficient* condition for ductile fracture is obtained when equation (3.9) is satisfied on a characteristic surface Σ, a characteristic-linked condition is not in itself *necessary* for ductile fracture and under certain conditions equation (3.9) may be satisfied first on a surface which does not coincide with Σ. Nevertheless, there will always be a stong tendency for ductile fracture-surfaces to initiate on characteristic surfaces Σ and this tendency is likely to increase with increasing constraint on the plastic-flow field.

3.4 The Plastic Limit-Load Model of Ductile Fracture and the Concept of a Dilational-Plastic Yield-Vertex Effect

It has been shown [7] that the critical condition (3.9) for ductile fracture by microvoid coalescence is fully equivalent to the attainment of a state where the sufficient condition for plastic stability (in the fundamental Kelvin-Dirichlet sense [10]) of a body containing microvoids is no longer satisfied; a result that follows directly from the first-order differences between a *virtual* mode of void coalescence and an *actual* homogeneous plastic-flow field [7]. We can therefore adopt the 'critical load' condition (3.9) for incipient microvoid coalescence knowing that this represents a mathematically rigorous condition for the end of stable plastic flow in the presence of microvoids.

For a three-dimensional plastic field, in a body containing microvoids, the current homogeneous macroscopic flow-field displays a *weak* dilational-plastic response resulting from the presence of approximately spherical microvoids and having an isotropic form (cf. Chapter 2.3 and Fig. 2.18). On the other hand, a virtual mode of incipient void coalescence, by localised internal necking (or plastic limit-load failure) across a sheet of microvoids, represents a *strong* dilational-plastic response [7], Fig. 3.9, which generally differs to first-order from the *weak* dilational-plastic response. The stability of the *weak* dilational-plastic response, in the potential presence of a *strong* dilational-plastic response, can be tested in the Kelvin-Dirichlet sense by the application of a virtual strain-rate field ($\dot{\epsilon}_1^{1c}$, 0, 0) representing the potential mode of void coalescence [7], Fig. 3.9. This yields the following sufficient condition for stability of the current *weak* dilational-plastic flow field [7]:

$$(\sigma_1^{1c} - \sigma_1) \; \dot{\epsilon}_1^{1c} > 0 \quad , \tag{3.11}$$

where σ_1 is the current maximum principal stress for the homogeneous flow field on the *weak* dilational-plastic yield surface and σ_1^{1c} is the stress on the *strong* dilational-plastic yield surface respresenting a virtual mode of incipient microvoid

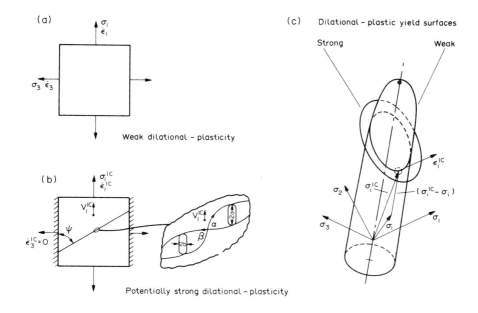

Fig. 3.9. The test of stability of a weak dilational-plastic field (a), in the presence of a potentially strong dilational-plastic response (b), by
the application of the virtual strain-rate field ($\dot{\epsilon}_1^{1c}$, 0, 0); also showing the principal stress-space representation (c) of the stability
condition (3.11). (After P.F. Thomason [7]).

coalescence by limit-load failure of the intervoid matrix. Clearly from the inequality
(3.11), stability of the *weak* dilational-plastic response is no longer guaranteed and
ductile fracture is imminent whenever:

$$(\sigma_1^{1c} - \sigma_1) \; \dot{\epsilon}_1^{1c} = 0 . \tag{3.12}$$

Now at incipient microvoid coalescence $\dot{\epsilon}_1^{1c} \neq 0$, and we obtain the critical load condition for ductile fracture:

$$\sigma_1^{1c} - \sigma_1 = 0 \quad , \tag{3.13}$$

which is completely equivalent to the two-dimensional result in equation (3.9) with
$\sigma_1^{1c} \equiv \sigma_n(k/k_n) (1 + \sqrt{V_f})^{-1}$.

The stress σ_1^{1c}, in the ductile-fracture criterion (3.13), is directly related to the plastic constraint factor $\sigma_n/2k_n$ at incipient limit-load failure of the intervoid matrix, and it is therefore necessary in developing the ductile-fracture model to obtain estimates of the constraint factors on a potential surface of microvoid coalescence. The full three-dimensional problem of obtaining the plastic limit-load condition for failure of the intervoid matrix across a sheet of ellipsoidal microvoids is far too complex to analyse and recourse must be made to approximate upper-bound solutions [11,12]; the full three-dimensional model for ductile fracture will be developed in Chapter 5. However, for the purpose of the present development of the two-dimensional ductile-fracture model, we now consider a relatively simple plane-strain analogue of localised plastic flow in a body containing microvoids, where slip-line field solutions are available for estimating the plastic limit-load state for the intervoid matrix [7]. The plane-strain plastic body (i.e., $\epsilon_2 = 0$) is assumed to contain a rectangular array of circular-cylindrical voids, in the initial state (Fig. 3.10(a)), where the array of voids is orientated at an angle ψ_o with respect to the X_1 direction of maximum principal stress. The initial dimensions of the voids in the principal directions X_1, X_3 are $2a_o$ and $2c_o$, respectively, and the initial inter-centre distances between adjacent voids is $2S_o$; the initial volume fraction of the cylindrical voids is therefore given by $V_f = \pi/4(a_o/S_o)^2$. When the principal axes of the strain-rates remain fixed in the X_1 and X_3 directions throughout the subsequent homogeneous plastic-flow process the current geometry of the microvoid array at a total plastic strain ϵ_1 will be as shown in Fig. 3.10(b), where the current spacing $2S$ of the void centre-lines along a potential surface of void coalescence ψ is given by $2S = 2S_o \exp(-\epsilon_1) \sin \psi_o/\sin \psi$, and ψ is given by $\cot \psi = \exp(2\epsilon_1) \cot \psi_o$ [7].

The initial state of strain $\epsilon_1 = 0$ in the present two-dimensional model (Fig. 3.10(a)) is broadly equivalent to the state existing at the microvoid-nucleation strain ϵ_1^n in a real material (cf. Section 3.1), we therefore use the microvoid nucleation state as the ground state and the current strain ϵ_1 is the *void-growth* strain which determines the ductile growth of the voids from the initial circular-cylindrical state to the current elliptical-cylindrical state (Fig. 3.10(b)). The ductile changes in shape and volume of the cylindrical voids can be represented by the principal dimensions (2a, 2c) of the resulting elliptical void-sections and the magnitudes of these parameters can be obtained on the asumption that they change in an identical manner to the in-plane dimensions of a spherical void in the Rice-Tracey model [13] for a state of plane-strain ($\nu = 0$) plastic flow (cf. Chapter 2.3 and Figs. 2.10, 2.12).

It is useful at this point to obtain some numerical estimates of the changing geometry of an initially square array of cylindrical holes, with increasing tensile strain ϵ_1 in a plane-strain plastic field, using the integrated form of the Rice-Tracey equations (2.16a and c) to give the principal dimensions 2a and 2c of the elliptical void section (i.e., taking $a/a_o \equiv R_1/R_o$ and $c/c_o \equiv R_3/R_o$ in equations (2.16)). This approach can be used to obtain estimates of the changing geometry (a/c, c/w) of a unit cell enclosing a void (Fig. 3.10), with increasing plastic strain ϵ_1, and these results [7] can then be combined to give a single parameter N representation of the

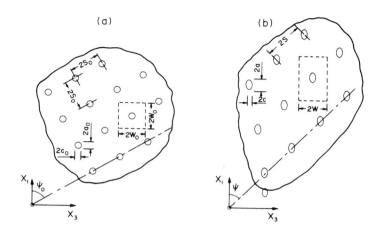

Fig. 3.10. The geometry and spacing of cylindrical voids in a plane-strain
plastic field showing (a) the initial undeformed state ($\epsilon_1 = \epsilon_1^n =$
0) and (b) the current deformed state ($\epsilon_1^F > \epsilon_1 > \epsilon_1^n = 0$) of the unit
cell. (After P.F. Thomason [7]).

current geometry of the intervoid matrix, where:

$$N = \frac{a}{w - c}.$$ \hfill (3.14)

A typical set of results for the variation of the neck-geometry parameter N, with
increasing plastic strain ϵ_1, is shown in Fig. 3.11 for the case where $\psi_o = \psi = \pi/2$;
these results are given for initial void volume fractions V_f of 0.005 and 0.02, at various constant mean-normal stresses σ_m in the range $4k \geq \sigma_m \geq k$, under both non-
hardening and linear-hardening conditions. The results in Fig. 3.11 show a clear
tendency for the neck-geometry parameter to approach a value $N \sim 1$, with increasing ϵ_1, and the effects are accentuated by increasing levels of mean-normal stress
σ_m and a reduction from linear-hardening to non-hardening plasticity.

The general problem of finding the plastic limit-load condition for incipient void
coalescence across a sheet of microvoids, at an arbitrary angle ψ to the direction of
maximum principal strain ϵ_1 (Fig. 3.10), is highly complex even for the present relatively simple plane-strain model, we therefore confine attention to the problem of
finding the plastic limit-load condition for a sheet of microvoids on a transverse
surface where $\psi = \psi_o = \pi/2$. At incipient limit-load failure of the intervoid matrix,
the plastic field will completely span the intervoid matrix and the plane-strain slip-
line fields will be statically determined by the zero-stress boundary conditions on
the surfaces of the adjacent elliptical holes, with the plastically deforming regions

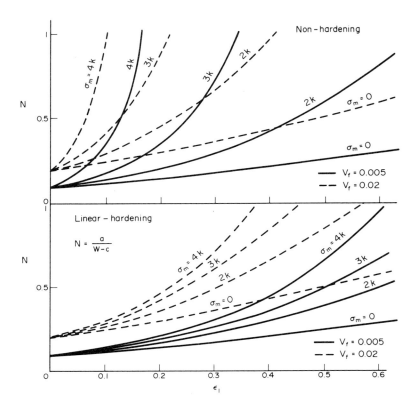

Fig. 3.11. The changes in the intervoid-geometry parameter N of the unit
cell, with increasing plastic strain ϵ_1, for various mean-normal
stresses σ_m and void volume fractions V_f, under both non-hardening and linear-hardening conditions.

enclosed by tangential velocity-discontinuities on the bounding slip lines that give
compatibility of deformation with the adjacent rigid regions, Fig. 3.12. The slip-line
field solutions for the incipient limit-load state between adjacent elliptical holes are
are available [7,14] and the results for the constraint factor $\sigma_1^{1c}/2k_n$ are given in Fig.
3.13 against c/w for various aspect ratios a/c of the holes. The results in Fig. 3.13 can
now be replotted against the neck-geometry parameter N and this has the interesting effect of 'collapsing' the data onto a single curve (Fig. 3.14).

For the present model of plane-strain plastic flow, in a body containing cylindrical
voids, the macroscopic principal stress $\sigma_1 = k + \sigma_m$ and the condition for incipient
ductile fracture by plastic limit-load failure of the intervoid matrix (equation (3.13))

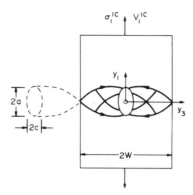

Fig. 3.12. The plane-strain slip-line fields, for virtual modes of void coalescence and incipient limit-load failure of the intervoid matrix, on a transverse surface $\psi = \pi/2$. Schematic representation showing the unit cell at an incipient fracture surface, with the deforming plastic zones enclosed by tangential velocity discontinuities. (After P.F. Thomason [7]).

can therefore be rewritten:

$$\sigma_1^{1c} = k + \sigma_m, \tag{3.15}$$

where k is related to k_n by equation (3.8). It is clear from Figs. 3.11 and 3.14 that in the initial stages of plastic flow ($\epsilon_1 \sim 0$) the neck-geometry parameter N is very small (N $\sim$ 0.1) and this corresponds to very large values of the plastic limit-load stress σ_1^{1c} ($\sim 6k_n$). In this case, $\sigma_1^{1c} > (k + \sigma_m)$ and the condition (3.11) for stable plastic flow in the porous solid is satisfied and ductile fracture cannot intervene. However, the results in Fig. 3.11 show that with increasing plastic strain ϵ_1, the neck-geometry parameter N increases rapidly, thus reducing the plastic limit-load stress σ_1^{1c} (Fig. 3.14) to the point where equation (3.15) is satisfied and ductile fracture by internal microscopic necking can commence across the "weakest" sheet of microvoids (cf. Fig. 3.6). Clearly, from the form of the incipient ductile-fracture condition (3.15) and the results in Fig. 3.11 and 3.14, an increase in both the volume fraction of voids V_f and the mean-normal stress σ_m will reduce the ductile-fracture strain; on the other hand, an increased work-hardening effect will reduce the growth rate of the N parameter and therefore tend to increase the fracture strain.

The results in Figs. 3.11 and 3.14 for ductile plastic flow and fracture in a porous solid, under plane-strain conditions, can be used to show [7] that the plastic limit-load mechanism for ductile fracture is conceptually equivalent to a sudden transformation from isotropic and weak dilational-plasticity to anisotropic and strong

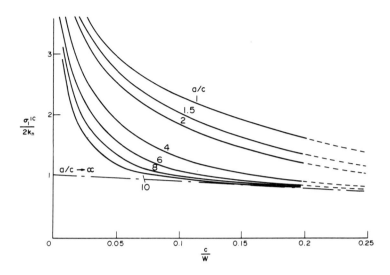

Fig. 3.13. The constraint factors $\sigma_1^{1c}/2k_n$ required to operate a virtual mode of void coalescence and initiate plastic limit-load failure between adjacent elliptical voids, for various unit-cell geometries. (After P.F. Thomason [7]).

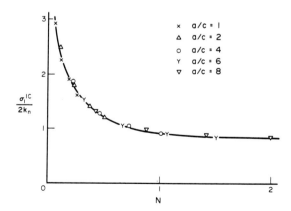

Fig, 3.14, The constraint factors $\sigma_1^{1c}/2k_n$ from Fig. 3.13 replotted against the neck-geometry parameter $N = a/(w - c)$.

dilational-plasticity; with the transformation in material response corresponding to the sudden development of a *dilational* yield vertex. The current mode of weak-dilational plastic flow in the plane-strain model is respresented by the stress vector σ_i ($\sigma_1 = \sigma_m + k$, $\sigma_2 = \sigma_m$, $\sigma_3 = \sigma_m - k$) which lies on the line $\mu = 0$ of the weak-dilational yield surface in principal stress space, Fig. 3.15, where μ is the Lode stress parameter [2]. Now the stress σ_1^{1c} for incipient plastic limit-load failure of the intervoid matrix is respresented by the strong dilational-yield surface $\sigma_1 = \sigma_1^{1c}$, lying parallel to the (σ_2, σ_3) plane (Fig. 3.15), and this intersects the line $\mu = 0$ at the point σ_i^{1c} corresponding to the stresses (σ_1^{1c}, $\sigma_2^{1c} \approx \sigma_1^{1c} - k_n$, $\sigma_3^{1c} \approx \sigma_1^{1c} - 2k_n$) on an element in a virtual mode of incipient limit-load failure of the intervoid matrix [7]. The strong dilational-yield surface $\sigma_1 = \sigma_1^{1c}$ has an associated dilational strain-rate vector $\dot{\epsilon}_i^{1c}$, lying normal to the surface and therefore parallel to the σ_1 stress axis, with components ($\dot{\epsilon}_1^{1c}$, 0, 0) which remain as *virtual* strain rates throughout the stable process of homogeneous ductile flow. The condition for stable ductile flow can therefore be rewritten:

$$(\sigma_i^{1c} - \sigma_i) \; \dot{\epsilon}_i^{1c} > 0 \quad , \qquad (3.16)$$

which is completely equivalent to condition (3.11) because $\dot{\epsilon}_2^{1c} = \dot{\epsilon}_3^{1c} = 0$ [7].

With increasing plastic strain ϵ_1 the neck-geometry parameter N increases continuously (Fig. 3.11(a)) and this leads to a continuous reduction in σ_1^{1c} (Fig. 3.14). The overall effect of increasing plastic strain ϵ_1 is a translation of the strong dilational-yield surface ($\sigma_1 = \sigma_1^{1c}$) towards the current active stress vector σ_i on the weak dilational-yield surface (Fig. 3.15(b)). Stable ductile flow will therefore be terminated when the strong dilational-yield surface translates to a position of contact with the active stress vector σ_i (Fig. 3.16), thus satisfying the ductile-fracture condition (3.13) (i.e., $\sigma_1^{1c} - \sigma_1 = 0$). At this point the *virtual* mode of void coalescence, represented by the strong dilational-plastic response, can become an *actual* mode and terminate the weak-dilational mode of plastic flow by the sudden intervention of an incipient mode of ductile fracture by void coalescence across a sheet of voids. The instant when the strong dilational-yield surface contacts the active stress vector σ_i on the weak-dilational yield surface represents the creation of a dilational-yield vertex (which is not to be confused with a deviatoric yield vertex, cf. Chapter 1.2) at which the strain-rate vector can suddenly switch from the weak-dilational state ($\dot{\epsilon}_1 + \dot{\epsilon}_2 + \dot{\epsilon}_3 \approx 0$) to the strong-dilational state ($\dot{\epsilon}_1^{1c} > 0$, $\dot{\epsilon}_2 = \dot{\epsilon}_3 = 0$) in the direction of the maximum principal-stress σ_1 axis. These effects are illustrated in Fig. 3.16 where it is to be noted that the weak dilational-yield surface remains smooth, with a unique normal strain-rate vector $\dot{\epsilon}_i$, up to the point of incipient ductile fracture, Fig. 3.16(b)). The dilational-yield vertex is therefore not visible in the π-plane projection (Fig. 3.16(b)) and is only apparent when the yield surface is viewed from a direction normal to the σ_m axis, Fig. 3.16(c).

The constructions in principal stress space (Figs. 3.15 and 3.16) were developed for the case of void coalescence on a transverse surface ($\psi = \psi_o = \pi/2$) and this led to

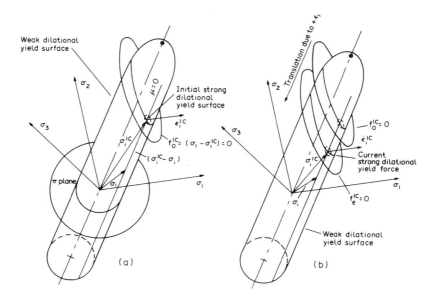

Fig. 3.15. (a) The representation in principal-stress space of the strong
and weak dilational-plastic yield surfaces, satisfying the sufficient condition for stability of weak-dilational plasticity. (b) The
translation of the strong dilational-plastic yield surface, due to
stable weak-dilational plastic flow, which will eventually lead to
the development of a dilational-yield vertex. (After P.F. Thomason [7]).

the unambiguous representation of incipient void coalescence by a single strong
dilational-yield surface $(\sigma_1 = \sigma_1^{1c})$. If, however, void coalescence were to develop
along a characteristic surface $\psi < \pi/2$ of the plastic velocity field (cf. Section 3.2) the
possibility arises that the exact value of σ_1^{1c} will depend on the orientation ψ of the
surface of void coalescence, and this would require multiple strong dilational-yield
surfaces to represent the potential limit-load failure of the intervoid matrix over a
range of possible ψ values. This would probably be a unnecessary complication of
the problem, however, because the upper-bound results for $\sigma_n \equiv \sigma_1^{1c}$ between an
array of square holes at various angular orientations ψ (Fig. 3.8) suggest that σ_1^{1c}
may well increase only slightly as ψ decreases from $\pi/2$ to $\pi/4$. It seems reasonable
therefore to take the strong dilational yield surface $(\sigma_1 = \sigma_1^{1c})$ for $\psi = \pi/2$ to represent the conditions for incipient ductile fracture on characteristic surfaces through-
out the range $\pi/2 \geq \psi \geq \pi/4$.

An important feature of the ductile-fracture model described above is that it recog-
nises the necessity to model the ductile fracture process in terms of a *dual* plastic

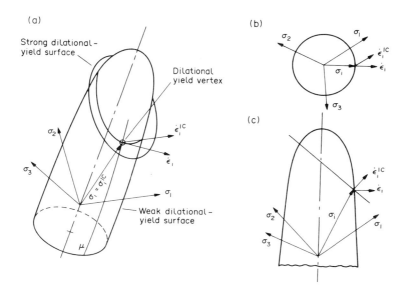

Fig. 3.16. (a) The representation in principal-stress space of the condition
for incipient ductile fracture when the translation of the strong
dilational-yield surface contacts the active stress point on the
weak dilational-yield surface and forms a dilational-yield vertex. Also showing two views of the dilational-yield vertex: (b)
along the mean-normal stress σ_m axis and (c) in a direction normal to the σ_m axis.

constitutive response for a body containing microvoids. When a porous solid is
modelled in terms of a *singular* or *unique* dilational-plastic response, the microvoids have effectively been removed from the body [15] to leave a completely
continuous solid with a *weak* dilational-plastic response similar to that resulting
from the presence of approximately spherical voids. Since the local plastic-stress
field σ_i^{1c} in a virtual mode of void coalescence is uniquely determined by the zero-
stress boundary conditions on the void surfaces and the relative velocity of the
rigid regions above and below the incipient fracture surface, it will generally differ
to first order from the current homogeneous stress field σ_i in a weak dilational-plastic solid [7]. In other words, we can draw the important conclusion that virtual
modes of void coalescence cannot begin as virtual modes of the current *weak* dilational-plastic state [7]. It follows from this that the full effect of the actual *presence*
of microvoids in a plastic *continuum* model can only be accounted for by incor-
porating a second *strong* dilational-plastic response to model a possible plastic
limit-load failure of the intervoid matrix. If a dual constitutive response is not used
in modelling the ductile-fracture process, then the primary mechanism of ductile

fracture by internal microscopic necking of the intervoid matrix has effectively been excluded as a controlling mechanism of fracture. This feature of the plastic limit-load model of ductile fracture [3,7] will be considered in more detail in Section 3.5, in relation to limitations on the validity of the Berg-Gurson model of ductile fracture.

It should be noted at this point that for the case of a work-hardening plastic solid, where k increases with total equivalent plastic strain $\bar{\epsilon}$ [2], the yield-shear stress k_n for the intervoid matrix increases in direct proportion to k and they continue to be related by equation (3.8). This is a direct consequence of the fact that the plastic limit-load velocity field is only a virtual (or potential) velocity field throughout the stable mode of ductile-flow where condition (3.11) remains satisfied; the limit-load velocity-field only transforms from a virtual to an actual velocity-field at the very instant of incipient ductile fracture when condition (3.13) is satisfied [7]. It follows from this that work-hardening effects can be readily included in the plastic limit-load conditions (equations (3.12), (3.13) and (3.16)) for ductile fracture by regarding both σ_i and σ_i^{1c} as increasing functions of the *current* macroscopic yield stress Y of the porous solid where, from equation (3.8), we have:

$$Y(\bar{\epsilon}) = \sqrt{3}\,k(\bar{\epsilon}) = \sqrt{3}\,(1 - V_f)k_n(\bar{\epsilon}) \quad . \tag{3.17}$$

Now both the virtual σ_i^{1c} and the actual σ_i stresses are directly proportional to $Y(\bar{\epsilon})$, so the critical conditions for ductile fracture by microvoid coalescence (equations (3.12), (3.13) and (3.16)) remain valid for work-hardening materials. This result is illustrated schematically in Fig. 3.17 where a state of isotropic hardening is assumed to occur homogeneously throughout the porous solid.

In principal stress space $(\sigma_1, \sigma_2, \sigma_3)$ (Fig. 3.17(a)) the work hardening effect leads to an increment $d\sigma_i$ from the initial active stress state σ_i on the weak dilational-yield surface and a corresponding increment $d\sigma_i^{1c}$ for the virtual stress vector σ_i^{1c} on the strong dilational-yield surface; the stress-increment vector $d\sigma_i^{1c}$ being the resultant of an increase due to work-hardening effects and a reduction due to the incremental change in the neck-geometry parameter N (Figs. 3.11 and 3.14). If the *current* stresses σ_i and σ_i^{1c} are normalised by the *current* yield stress Y at all stages of ductile plastic flow, the results for work-hardening conditions can be represented by single strong and weak dilational-yield surfaces as shown in Fig. 3.17(b), thus confirming the general validity under work hardening conditions of the condition (3.16) for stable ductile flow in a porous solid, and the critical condition (3.13) for ductile fracture. It should be noted that this does not imply that ductile fracture conditions are independent of work-hardening effects; the results in Fig. 3.11 show that work-hardening effects reduce the growth rate in the neck-geometry parameter N thus reducing the rate of reduction in σ_1^{1c} (Fig. 3.14) with increasing plastic strain. Hence, under work-hardening conditions the dilational-yield surface ($\sigma_1 = \sigma_1^{1c}$) (Fig. 3.17(b)) translates towards the active stress point σ_i at a much slower rate

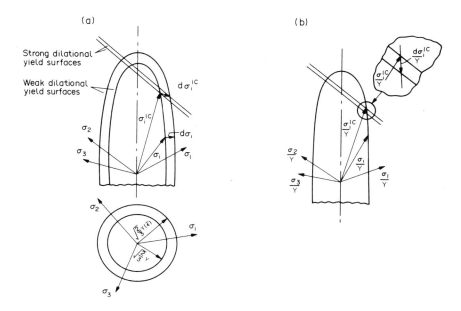

Fig. 3.17. (a) A schematic representation of the effect of isotropic work-hardening on the strong- and weak-dilational yield surfaces and (b) the equivalence to the non-hardening state when the stresses are normalised by the current uniaxial yield stress $Y(\bar{\epsilon})$.

than for non-hardening plastic flow. We can now generalise the conditions for both stable ductile flow (conditions (3.11) and 3.16)) and incipient ductile fracture (equation (3.13)) to account for work-hardening effects in ductile porous solids, by rewriting the respective ductility conditions in the following forms:

$$(\sigma_i^{1c}(\bar{\epsilon}) - \sigma_i(\bar{\epsilon}))\,\dot{\epsilon}_i^{1c} > 0 \, , \tag{3.18}$$

for stable ductile flow and:

$$\sigma_i^{1c}(\bar{\epsilon}) - \sigma_i(\bar{\epsilon}) = 0 \, , \tag{3.19}$$

for incipient ductile fracture. In the above generalised ductile-fracture conditions the plastic limit-load stress $\sigma_1^{1c}(\bar{\epsilon})$ can be evaluated approximately from equation (3.17) and the non-hardening value of $\sigma_1^{1c}/2k_n$ by the relation:

$$\sigma_1^{1c}(\bar{\epsilon}) = \frac{\sigma_1^{1c}}{2k_n}\,2k_n(\bar{\epsilon}) = \frac{\sigma_1^{1c}}{2k_n} \cdot \frac{2}{\sqrt{3}}\,\frac{Y(\bar{\epsilon})}{(1 - V_f)} \, ; \tag{3.20}$$

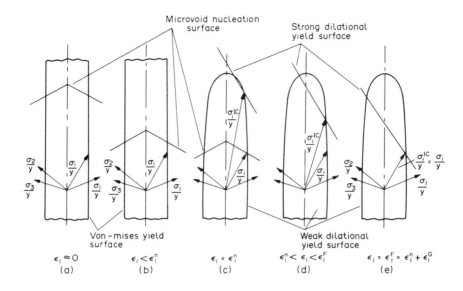

Fig. 3.18. The complete process of ductile fracture in metals shown
schematically in normalised principal-stress space.

this relation is based on the assumption that the state of hardening in the intervoid matrix is approximately homogeneous up to the point of incipient microvoid coalescence. The application of void-coalescence criteria to the problem of estimating the void-growth strain ϵ_1^G at ductile fracture in alloy/particle systems will be considered in Chapter 5.

It is now possible to construct an interesting representation in principal stress space (σ_1, σ_2, σ_3) of the complete process of ductile fracture in work-hardening solids from the initial state where microvoids do not exist, through the microvoid nucleation and growth states to the point of incipient ductile fracture by plastic limit-load failure of the intervoid matrix, Fig. 3.18. In the initial undeformed plastic state (Fig. 3.18(a)) the microstructural inclusions and second-phase particles are firmly bonded to the matrix and microvoids are not present; a dilational-yield surface therefore does not exist and the yield surface has a purely deviatoric (von Mises) form (cf. Chapter 1.2). However, a conical surface, representing the criterion for microvoid nucleation by matrix/particle debonding (cf. Chapter 2.2), exists well above the active stress vector σ_i (Fig. 3.18(a)) and with work-hardening plastic flow this conical surface will effectively translate towards the active stress vector because the principal stresses are normalised by the current macroscopic yield stress $Y(\bar{\epsilon})$ (Fig. 3.18(b)). This process continues until the conical void-nucleation surface comes into contact with the active stress vector σ_i (Fig. 3.18(c)), whereupon small and widely separated microvoids are nucleated (i.e., $\epsilon_1 \equiv \epsilon_1^n$) transforming the deviatoric yield surface into a *weak* dilational-yield surface and bringing into existence a virtual *strong* dilational-yield surface well above the active stress vector σ_i (Fig. 3.18(c)); at this point, the porous solid has a *dual* plastic constitutive response with an actual weak-dilational response and a potential (or virtual) strong-dilational response. With continuing plastic strain beyond the void-nucleation strain ϵ_1^n microvoid growth causes the translation of the strong dilational-yield surface towards the active stress vector (Fig. 3.18(d)) and ductile plastic flow is terminated in the porous work-hardening solid when the strong dilational-yield surface becomes coincident with the active stress vector σ_i on the weak dilational-yield surface (Fig. 3.18(e)); thus forming a dilational-yield vertex (Fig. 3.16) at the incipient ductile-fracture strain $\epsilon_1^F = \epsilon_1^n + \epsilon_1^G$.

3.5 The Berg-Gurson Model of Ductile Fracture by Strain Localisation in Homogeneous Dilational-Plastic Flow Fields.

An alternative model of ductile fracture, which differs fundamentally from the plastic limit-load model described in the Section 3.4, was originally proposed by Berg [16] and subsequently developed in the work of Gurson [17], Rudnicki and Rice [18], Yamamoto [19], and Tvergaard [20]. This model is based on the assumption that the complete ductile-fracture process of void growth and coalescence can be described entirely in terms of the *weak* dilational-plastic response of an elastic/plastic continuum, which displays the macroscopic effects that would result from a distribution of spherical microvoids but does not actually *contain* microvoids (cf. Chapter 2.3) [7,15]. In this approach, ductile fracture is regarded as the result of an instability in the weak dilational-plastic flow field allowing the formation of an intense localised 'shear' band. It is further assumed that void coalescence by internal necking of the intervoid matrix is a secondary effect which can only develop *after* the incipient formation of an intense 'shear' band. Hence, the strong dilational-plastic response associated with plastic limit-load failure of the intervoid matrix, which cannot occur in the Berg-Gurson model due to the effective removal of microvoids [15], is neglected and the primary mechanism of ductile fracture has therefore been inadvertently eliminated from the model. As will be shown later, it is this failure to appreciate the dominant effect of internal microscopic necking by plastic limit-load failure of the intervoid matrix, which can only be modelled adequately by the incorporation of a (dual) strong dilational-plastic response [7,15], that severely limits the validity of the Berg-Gurson model.

The weak dilational-plastic response of a body containing microvoids was modelled in the work of Gurson [17] by a unit spherical cell with a single central void and the corresponding dilational-yield surface for the aggregate was found to have the form:

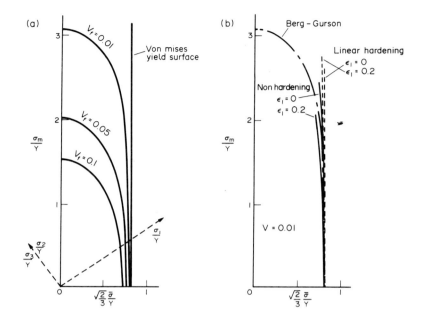

Fig. 3.19. (a) The form of the weak dilational-plastic yield surfaces from the Berg-Gurson model for various volume-fraction of voids V_f. (b) A comparison of the weak dilational-plastic yield loci obtained from the Rice-Tracey model (Fig. 2.18) and the corresponding Berg-Gurson yield loci.

$$\Phi = \frac{3}{2} \frac{\acute{\sigma}_{ij}\acute{\sigma}_{ij}}{Y_n^2} + 2V_f \cosh\left(\frac{3}{2} \frac{\sigma_m}{Y_n}\right) - (1 + V_f^2) = 0, \tag{3.21}$$

where $\acute{\sigma}_{ij} = \sigma_{ij} - \delta_{ij}\sigma_m$ are the macroscopic deviatoric stress components, Y_n is the uniaxial yield stress of the matrix and V_f is the volume fraction of voids. Where it is to be noted that Φ reduces to the von Mises isotropic-hardening form when $V_f = 0$. Typical results obtained from equation (3.21) are shown in Fig. 3.19(a) for values of void volume-fraction in the range $0.01 \leq V_f \leq 0.1$. It is interesting at this point to compare the results from the Berg-Gurson model for $V_f = 0.01$ (Fig. 3.19) with the dilational-plastic results obtained in Chapter 2.3 for $V_f = 0.01$, using the Rice-Tracey model of void growth [13]. The two sets of results for the weak dilational-plastic yield loci are compared in Fig. 3.19(b) where there is seen to be good agreement between the non-hardening results from the Rice-Tracey model and those from the Berg-Gurson model; even though the two models for the plastic flow fields are known to differ substantially. This good agreement between the two sets of results in Fig. 3.19 indirectly confirms the validity of the simple 'law-of-mixtures' relation between the microscopic (k_n) and macroscopic (k) values of yield shear stress (equation (3.8)) for relatively low volume fractions V_f of voids and low σ_m stresses.

The conditions under which a *continuum* model, with a weak dilational-plastic response of the form given in equation (3.21), will first exhibit a localisation of deformation into an intense flow band or 'shear' band were presented by Rudnicki and Rice [18] and applied in detail to the problem of ductile fracture by Yamamoto [19]. The localisation or bifurcation condition was based on the assumption that the constitutive response of a solid containing microvoids would have a unique constitutive rate relation of the form:

$$\overset{\triangledown}{\sigma}_{ij} = L_{ijkl} D_{kl} \; ; \tag{3.22}$$

where $\overset{\triangledown}{\sigma}_{ij}$ is the Jaumann co-rotational stress rate [21], L_{ijkl} is the modulus tensor and D_{kl} is the rate of deformation. If there is a unique relation of the form given by equation (3.22) then L_{ijkl} remains the same inside and outside a band of incipient bifurcation and it can be shown that [18,19] for incipient bifurcation on a band with a normal in the X_2 direction:

$$(L_{2jk2} - R_{jk}) q_k = 0 \quad , \quad j = 1,2,3 \quad , \tag{3.23}$$

where R_{jk} is a modulus tensor related to the co-rotational stress rate and $q_k(X_2)$ represents the non-uniformity of deformation in a 'shear' band (i.e., $\Delta(\partial V_i/\partial x_j) = q_i(X_2)\delta_{j2}$, $(i,j = 1,2,3)$). The condition for bifurcation in a solid with a weak dilational-plastic response is therefore reached when this equation first allows non-zero solutions for q_k and this is given by [18,19]:

$$\det \left| L_{2jk2} - R_{jk} \right| = 0 . \tag{3.24}$$

It is at this point that any possibility of the Berg-Gurson model including the strong destabilising effect of a plastic limit-load failure of the intervoid matrix (i.e., strong dilational plasticity), as a controlling mechanism of ductile fracture, has finally been eliminated. Equation (3.24) is formulated in terms only of second-order variations from a state of weak-dilational plasticity in a solid that has had the microvoids effectively removed (the field equations being satisfied at *all* points, both inside and outside the deformation band [18]). It is perhaps unfortunate that the bifurcation condition (3.24) was derived for problems of instability in rock and soil mechanics, where the tensile response differs markedly from that in metals, and was never reformulated to take account of the fundamentally important fact that metallic ductility can be governed by the condition of plastic limit-load failure of the intervoid matrix. The modifications that are necessary to condition (3.23), to allow for metallic ductility effects are discussed later. A further unfortunate consequence of the unmodified application of bifurcation conditions, which were primarily developed in terms of rock and soil mechanics [18], is the use of the term 'localised shear band' which implies that the tangential discontinuity Δ_T in the velocity gradients, across an incipient fracture surface, will greatly exceed the normal discontinuity Δ_N. In rock- and soil-mechanics problems such shear bands can develop, but in tensile ductility problems for metals, Δ_T and Δ_N on a surface of incipient void coalescence ψ must be related by the compatibility condition (3.4). In metallic ductility problems the terms 'shear band' and 'shear fracture' should therefore be reserved for those special cases where $\Delta_T \gg \Delta_N$; in the general case 'shear band' should be replaced by 'deformation band' (cf. Section 3.2 and Chapters 6.5, 7.3 and 8.2).

Using the bifurcation condition (3.24) for flow localisation in a Berg-Gurson solid continuum, with the weak dilational-plastic yield function (3.21), it can be shown [18,19] that the critical microscopic hardening modulus h_{mcr} at incipient flow localisation is given by:

$$h_{mcr} = \frac{3(1 - V_f)}{\left(w + \left(\frac{3\sigma_m}{Y_n}\right)\alpha\right)^2}\left\{-\frac{3G}{2}(1 + v)\left(\frac{\sigma'_{II}}{Y_n} + \frac{2\alpha}{3}\right)^2 + Y_n(1 - V_f)\alpha\left[\cosh\left(\frac{3}{2}\frac{\sigma_m}{Y_n}\right) - V_f\right]\right\};$$

(3.25)

where G is the elastic shear modulus, v is Poisson's ratio, $\alpha = \frac{1}{2}V_f\sinh(3\sigma_m/2Y_n)$, $w = [(1 + V_f^2) - 2V_f\cosh(3\sigma_m/2Y_n)]$, $\sigma'_I, \sigma'_{II}, \sigma'_{III}$ are the macroscopic deviatoric principal stresses, and the critical angle θ_{cr} of the deformation band with respect to the σ_{III} direction, which maximises h_m, is given by:

$$\cos 2\theta_{cr} = \frac{-(1 - 2v)(\sigma'_{II}/Y_n) + (4(1 + v)/3)\alpha}{(\sigma'_I - \sigma'_{III})Y_n}.$$

(3.26)

If the matrix material is now assumed to obey an elastic/plastic power-hardening law of the form $\tau/\tau_y = \gamma/\gamma_y$ for $\tau < \tau_y$ and $\tau/\tau_y = (\gamma/\gamma_y)^n$ for $\tau \geq \tau_y$ (where τ and γ are the shear stress and shear strain, respectively, and τ_y, γ_y are the yield point values) the microscopic hardening modulus h_m can be written in the form:

$$h_m = \frac{3Gn}{\left(\frac{G\bar{\epsilon}}{\sigma_y}\right)^{1-n} - n};$$

(3.27)

where $\sigma_y = \sqrt{3}\,\tau_y$, n is a strain-hardening exponent, and $\bar{\epsilon}$ is the equivalent strain. The critical strain ϵ_I at incipient flow localisation into a deformation band is therefore obtained by substituting h_{mcr} from equation (3.25) for h_m in equation (3.27).

The limitations of a model of ductile fracture based on the second-order bifurcation condition (3.24) for a Berg-Gurson solid, showing only a weak dilational-plastic response, is readily illustrated with the aid of equation (3.25) by finding h_{mcr} for a number of basic stress systems. As pointed out by Yamamoto [19], in a state of uniaxial tension (σ_I, $\sigma_{II} = \sigma_{III} = 0$), with volume fractions of voids in the range $0 < V_f < 0.1$, equation (3.25) reduces to the highly negative form:

$$h_{mcr} \approx -\frac{(1 + \nu)G}{2(1 - V_f)^3} \tag{3.28}$$

and there is no tendency for deformation-band localisation at finite strain. Similar effects occur for a state of pure shear (σ_I, $\sigma_{II} = 0$, $\sigma_{III} = -\sigma_I$) where equation (3.25) reduces to the form $h_{mcr} = 0$; thus requiring incredibly large strains for localisation. In both the uniaxial-tension and pure-shear flow fields the limit-load model of ductile fracture (Section 3.4), which represents a dual constitutive response involving both strong and weak dilational-plastic constitutive relations, gives ductile fracture strains of the correct order (i.e., $\epsilon_I^F < 1$) [3,7,15,22].

Under plane-strain tension conditions (σ_I, $\sigma_{II} = \sigma_I/2$, $\sigma_{III} = 0$), equation (3.25) can be rewritten to a close approximation, for void volume-fractions in the range $0 < V_f < 0.1$, in the form:

$$h_{mcr} \approx \frac{0.68G\,V_f}{(1 - V_f)} \left[\frac{3Y_n}{G} - (1 + \nu)\,\frac{0.7\,V_f}{(1 - V_f)^2} \right]. \tag{3.29}$$

For typical values of the parameters in this equation h_{mcr} is positive for $V_f < 0.05$ but becomes strongly negative with further increase in V_f. This leads to the absurd result that large void volume-fractions ($V_f > 0.05$) will give very much higher ductility than low volume fractions ($V_f \sim 0.03$). The initial localisation strains for plane-strain tension with V_f in the range $0 < V_f < 0.05$, obtained by Yamamoto [19], are shown in Fig. 3.20 for strain hardening exponents in the range $0.05 \leq n \leq 0.2$. Although these results are more realistic than the virtually infinite critical strains in uniaxial tension and pure shear, they are still very much larger than those expected in real materials [15]. The results from the plastic limit-load model of ductile fracture (Section 3.4), under similar states of plane-strain tension [3], are also plotted for comparison in Fig. 3.20 and it is clear that they are more physically-realistic results, being about an order of magnitude smaller and implying that the condition for plastic limit-load failure of the intervoid matrix is reached well before the condition for deformation-band localisation in a solid displaying only a weak dilational-plastic response. As pointed out previously [7,15] plastic limit-load failure of the intervoid matrix cannot occur with the Berg-Gurson model because the microvoids have been 'removed' and are represented only by their effect in giving the *weak* dilational-plastic response (equation (3.21)). The resulting dilational-plastic solid is therefore completely continuous with the stress- and velocity-field equations existing

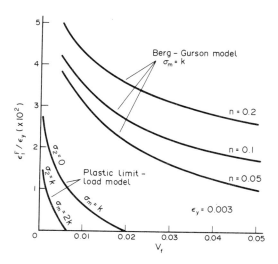

Fig. 3.20. The normalised critical localisation strains for various values of
the strain-hardening exponent n and void volume fraction V_f in
plane-strain tension; from Yamamoto's results for the Berg-
Gurson model [19]. Also showing the corresponding critical
ductile-fracture strains from the plastic limit-load model of ductile fracture [3]. (After H. Yamamoto, Int. J. Fracture, 1978, 14,
p.347 and P.F. Thomason, J. Inst. Metals, 1968, 96, p.360).

at all points both inside and outside the deformation bands [18,19]; there are therefore no voids with stress-free surfaces within the deformation bands of the continuum model and plastic limit-load failure of the intervoid matrix is therefore inadvertently excluded as a controlling mechanism of ductile fracture. It is for this
reason that the Berg-Gurson model can operate at strains well in excess of those
that lead to ductile fracture by internal microscopic necking (i.e., plastic limit-load
failure) of the intervoid matrix in real materials; consequently the Berg-Gurson
model gives a totally unrealistic description of the ductile fracture process in metals.

The primary limitation on the validity of the Berg-Gurson model can now be illustrated by a comparison of the bifurcation conditions for deformation-band localisation (equation (3.23)) and for plastic limit-load failure of the intervoid matrix (equation (3.12)). With a plastic limit-load model of ductile fracture [7], it is recognised
that, in the initial stages of ductile void-growth, the internal-necking mode of void
coalescence represents a *virtual* mode of strong dilational plasticity ($\sigma_i - \sigma_i^{1c} = 0$)
with associated stresses $\sigma_i \equiv (\sigma_1^{1c}, \sigma_2^{1c}, \sigma_3^{1c})$ which differ to first order from the current state of weak dilational plasticity $\sigma_i \equiv (\sigma_1, \sigma_2, \sigma_3,)$ on the Berg-Gurson yield

surface $\Phi = 0$. This fundamental result follows from the fact [7] that the state of incipient plastic limit-load failure of the intervoid matrix is uniquely determined by the zero-stress boundary conditions on the void surfaces and the displacement of the adjacent rigid regions above and below the surfaces of void coalescence (i.e., the Cauchy problem in plane-strain plasticity [2,7,14]). The *virtual* strong dilational-plastic response therefore differs to first order from the *actual* weak dilational-plastic response and thus remains a virtual mode right up to the point of incipient void coalescence, whereupon it can transform into an actual mode. The ductile-fracture process must therefore be modelled in terms of a dual (i.e., strong and weak) dilational-plastic response because the internal-necking mechanism of ductile fracture is distinct and cannot begin as a virtual mode of the weak dilational-plastic response $\Phi = 0$ [7]. This fundamental difference between the bifurcation conditions from the plastic limit-load model and the Berg-Gurson model can be seen by rewriting the stresses in equation (3.12) in terms of both their first order and second order components, i.e., $(\sigma_1^{1c} + \dot{\sigma}_1^{1c} - \sigma_1 - \dot{\sigma}_1)\dot{\epsilon}_1^{1c} = 0$. Now $\dot{\epsilon}_1^{1c}$ is proportional to the q_k terms in equation (3.23) so that (3.12) can be rewritten in the form:

$$(\sigma_1^{1c} - \sigma_1 + \dot{\sigma}_1^{1c} - \dot{\sigma}_1)\, q_k = 0 \quad . \tag{3.30}$$

Comparing this equation with the conditions for incipient deformation-band localisation in the Berg-Gurson model (equations (3.23) and (3.24)) it is clear that the first order difference $(\sigma_1^{1c} - \sigma_1) \gtrless 0$ has been neglected in formulating the bifurcation conditions and only the second-order (stress rate) terms $(\dot{\sigma}_1^{1c} - \dot{\sigma}_1)$ have been retained, which are approximately related by the stress-rate constitutive law (equation (3.22)) to $(L_{2jk2} - R_{jk})$ in both equation (3.23) and the bifurcation condition (3.24). This is the reason for the Berg-Gurson model showing a strong resistance to deformation-band localisation at realistic ductile-fracture strains and the extreme sensitivity of the final localisation strains to such second-order effects as the precise gradient of a 'saturated' work-hardening yield-stress curve and the ratios of the stresses to the material elastic moduli [18,19].

In an attempt to overcome the obvious limitations on the validity of a model formulated only in terms of second-order effects of a weak dilational-plastic response, it was proposed that the model could be made more realistic if it was assumed that the ductility of real materials was controlled primarily by the presence of "initial imperfections" [19]. It was found, however, that the initial imperfections would need to have a volume fraction of voids five times greater than that of the material outside the imperfection if realistic fracture strains were to be achieved [19]. It is of course absurd to propose that the normally observed ductility of structural materials is primarily dependent on the presence of imperfections of such severity. If this were the case and the Berg-Gurson model were valid then metallurgical processing could readily be devised to eliminate these severe imperfections and then give the apparently unlimited ductility predicted by the model for many stress states. This of course would be wasted effort: the Berg-Gurson model neglects the

controlling effect of plastic limit-load-failure of the intervoid matrix on ductile fracture, which brings about fracture at relatively low strains without the need for "initial imperfections". This does not of course mean that structural materials do not
contain initial imperfections; it does mean, however, that when they are present in
a severity five times greater than the normal volume fraction of voids, plastic limit-
load failure will occur much more readily in these regions thus giving very severely
reduced ductility.

An alternative approach to the problem of attempting to validify the Berg-Gurson
model of ductile fracture has been made by Tvergaard [20] and although this is less
objectionable on physical grounds than the "initial imperfection" approach, it is
nevertheless of doubtful validity as a model for the micro-mechanics and physics
of ductile fracture processes. Tvergaard's modification to the Berg-Gurson model
consists of rewriting equation (3.21) in terms of a set of arbitrary adjustable
parameters (q_1, q_2) in the form:

$$\Phi = \frac{3}{2} \frac{\sigma'_{ij} \, \sigma'_{ij}}{Y_n^2} + 2(q_1 V_f) \, \cosh\left(\frac{3 q_2 \sigma_m}{2 Y_n}\right) - (1 + (q_1 V_f)^2) = 0 \; ; \qquad (3.31)$$

where $q_1 > 1$, $q_2 > 1$. It is clear from the form of equation (3.31) that the parameters
q_1 and q_2 merely amplify the actual values of V_f and σ_m to the artificially elevated
levels $q_1 V_f$ and $q_2 \sigma_m$, respectively. Hence, if the q_i values ($q_1 = 2$, $q_2 = 1$) are required
to give realistic ductile-fracture strains for a material with an actual volume fraction
of microvoids $V_f = 0.05$, than a dilational yield surface corresponding to $V_f = 0.1$ has
actually been used in the analysis. Conversely, if the q_i values ($q_1 = 1$, $q_2 = 2$) are
required to give realistic fracture strains at a mean-normal stress $\sigma_m/Y_n = 2$, then an
artificially elevated mean-normal stress $\sigma_m/Y_n = 4$ has actually been used in the
analysis. The general effect of increasing the adjustable parameters (q_1, q_2) is illustrated in Fig. 3.21, where it is clear that quite small increases in q_i can lead to very
substantial modifications to the form of the 'weak' dilational-yield surface. Such a
drastic distortion of the basic constitutive model of a porous solid in order to obtain
'realistic' ductile fracture strains can hardly be regarded as giving an insight into
the basic mechanisms of ductile fracture.

The main objection to Tvergaard's approach is that he is effectively distorting the
weak dilational response (equation (3.21)) into a strong and *actual* dilational
response (equation (3.31)) at the very onset of the ductile void-growth phase when
the strong dilational response should still be only a *virtual* mode. In fact, the finite-
element model used by Tvergaard to 'calibrate' the modified Berg-Gurson yield
surface was effectively an incipient mode of void coalescence in an elastic/plastic
work-hardening solid [20], analogous to the rigid/plastic limit-load mode shown in
Fig. 3.12. The distortion of the strong and weak dilational-plastic yield surfaces,
involved in Tvergaard's modification to the Berg-Gurson model, is illustrated in
Fig. 3.22. The strain dependence [7] of the strong dilational-plastic response leads

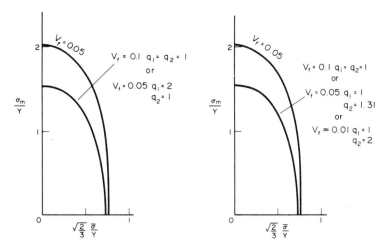

Fig. 3.21. The effect of changes in the adjustable parameters (q_1, q_2) in Tvergaard's modification [20] to the Berg-Gurson yield surface.

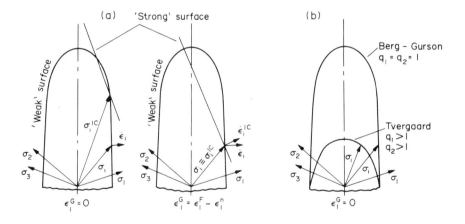

Fig. 3.22. A schematic illustration of the way in which Tvergaard's adjustable parameters (q_1, q_2) circumvent the gradual approach of the transition from weak- to strong-dilational plasticity (a), by replacing the weak dilational-plastic response with a strong dilational-plastic response (b) at the very onset of the microvoid-growth phase $(\epsilon_1^G = 0)$.

to a *progressive* truncation of the weak dilational-yield surface [7,15] (Fig. 3.22(a)), but Tvergaard's modification causes an *immediate* distortion of the weak dilational-yield surface to bring it into close proximity with the final truncating surface of strong dilational plasticity (Fig. 3.22(b)). However, the distorted Berg-Gurson yield surface is not a "good fit" to the final strong dilational-yield surface and the whole of this severe distortion has occurred at the very onset of void growth (i.e., $\epsilon_1^G = 0$); this leads to the Tvergaard model giving results that are statically inadmissible at the microstructural level [12] with the intervoid matrix being unable to support the equilibrium loads if the voids are 'returned' to the model.

An alternative modification to the Berg-Gurson model, in the spirit of Tvergaard's approach, which would remove some of the above objections and make the model very similar to the plastic limit-load model (Section 3.4) would be to replace the constant q_i parameters by a new set of parameters $p_1(\bar{\epsilon})$, $p_2(\bar{\epsilon})$ which are equal to unity at an equivalent plastic strain $\bar{\epsilon} = \epsilon_1^G = 0$ and increase monotonically with equivalent plastic strain up to the point of ductile fracture; the Berg-Gurson yield surface would then be represented by:

$$\Phi = \frac{3}{2} \frac{\sigma_{ij}'\sigma_{ij}'}{Y_n^2} + 2\,(p_1(\bar{\epsilon})V_f)\ \cosh\left(\frac{3\,p_2(\bar{\epsilon})V_f}{2\ Y_n}\right) - (1 - (p_1(\bar{\epsilon})V_f)^2) = 0 \ . \quad (3.32)$$

In this alternative modified form the Berg-Gurson model would compare schematically with the plastic limit-load model as shown in Fig. 3.23. It should be noted, however, that this Berg-Gurson model would still retain the symmetrical form of dilational-yield surface (about the mean normal stress σ_m axis) appropriate to a model containing spherical voids; only the plastic limit-load model has the strong dilational-plastic response of unsymmetrical form corresponding to the effects of elongated hole-growth in the maximum strain ϵ_1 direction [7,15].

3.6   Yield-Vertex and Kinematic-Hardening Models of Ductile Fracture

In further attempts to overcome the strong resistance to deformation-band localisation [18,19] in a dilational plastic continuum which does not actually contain microvoids (and therefore can exhibit only a weak dilational response), it has been proposed [18,23,24] that the development of a 'deviatoric' yield vertex would have a strong destablising effect on the condition for localisation into a deformation band (equations (3.23) and (3.24)) and might therefore be the explanation for the wide descrepancy between the very large predicted fracture strains from the Berg-Gurson model and the very much smaller ductile-fracture strains measured in corresponding experiments. A somewhat similar attempt to eliminate these descrepancies has also been based on the destabilising effects that would accompany the increased yield surface curvature [25,26] resulting from a material showing predominantly kinematic (or translational) hardening as distinct from isotropic

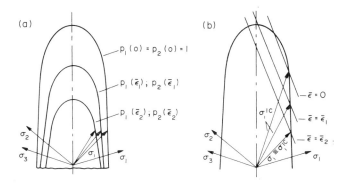

Fig. 3.23. (a) A schematic illustration of the effect of the $p_1(\bar{\epsilon})$ and $p_2(\bar{\epsilon})$ parameters on form of the Berg-Gurson yield surfaces with increasing plastic strain $\bar{\epsilon}$. (b) The corresponding effects of plastic strain on the strong- and weak-dilational yield surfaces in the plastic limit-load model of ductile fracture.

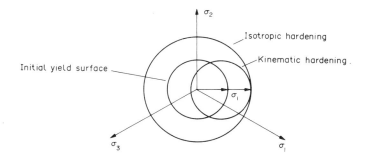

Fig. 3.24. The increasing yield-surface curvature resulting from kinematic hardening in comparison to isotropic hardening.

hardening, Fig. 3.24 (cf. Chapter 1.3). These additional attempts to validate the Berg-Gurson model, which still lead to predictions of ductile-fracture strains well in excess of both those observed in practice [18,23-26] and those predicted by the plastic limit-load model [3,7,11,12,15], stand or fall on the results from experimental and theoretical determinations of the form of yield and loading surfaces (cf. Chapter 1.3). Both experimental results [27] and physical theories of plasticity [28] show that if a deviatoric yield vertex exists at all it is erased by a plastic strain-increment of the order of $0.1\mu\epsilon$. In addition, experimental results show that the departure from isotropic hardening is quite small [29], depending critically on the magnitude of the plastic strain-increment used to define the yield surface (cf. Chapter 1.3); the larger this plastic strain-increment the closer the loading surface to an isotropic-hardening form. There is therefore no sound experimental basis to support the validity of either the yield-vertex or kinematic-hardening modifications to the Berg-Gurson model of ductile fracture. In fact, the yield-vertex models [18,23,24] are formulated in terms of a parameter, the vertex modulus, which would be virtually impossible to measure by any feasible experimental technique. The kinematic hardening models [25,26] also have the disturbing feature that, in terms of a deformation-band localisation mechanism, they imply a certain degree of 'memory' of the previous history of plastic flow.

The fundamental reason for the failure of the Berg-Gurson model to accurately describe the ductile-fracture process in metals lies in the fact that the microvoids have been effectively removed from the elastic/plastic continuum and are represented only by the *effect* that spherical voids would have in giving a weak dilational-plastic response. The bifurcation conditions are consequently formulated only in terms of the second-order stress-rates; the weak-dilational stresses inside the outside a deformation band being always in equilibrium and equal to first-order when microvoids are no longer present. When microvoids are actually present in the model, the extensional void growth in the direction of maximum principal strain ϵ_1 has the effect of introducing a possible (dual) strong dilational-plastic response which initially differs to first-order from the weak-dilational response and therefore remains a virtual response until the increasing void-elongation brings about an equality of the strong- and weak-dilational stresses. It is this failure of the Berg-Gurson model to account for the dual-constitutive nature of a plastic body that actually contains microvoids, which leads to its failure to accurately model the ductile fracture process. Ingenious modifications to the form of the weak-dilational constitutive response, as included in the yield-vertex and kinematic-hardening models, cannot change the basic error of the Berg-Gurson approach, where the microvoids are effectively removed from the elastic/plastic continuum and the possiblity of internal microscopic necking by plastic limit-load failure of the intervoid matrix has thus been inadvertently excluded.

However, a simple modification can be made to the Berg-Gurson model, which makes it fully consistent with the plastic limit-load model of ductile fracture (Section 3.3), and this consists of treating the Berg-Gurson yield surface (equation (3.21)) as the weak-dilational response in a dual constitutive model. The strong dilational-yield surface corresponding to the effects of a potential (virtual) mode of void coalescence, by plastic limit-load failure of the intervoid matrix, would then progressively truncate the 'weak' Berg-Gurson yield surface (Fig. 3.23(a)) [7,15] with increasing plastic strain. The critical condition for incipient ductile fracture would then be given by the first-order bifurcation conditions (3.12) and (3.13) in place of the second-order conditions (3.23) and (3.24). This modification to the Berg-Gurson model would have a more sound theoretical basis than that proposed in equation (3.31), and could be applied more readily in analytical modelling of ductile-fracture processes.

REFERENCES

1. Hill, R. *Discontinuity Relations in Mechanics of Solids,* Chapter VI of *Progress in Solid Mechanics - Volume II,* edited by I.N Sneddon and R. Hill, p.247, North-Holland Publishing Co., Amsterdam, 1961.
2. Hill, R. *The Mathematical Theory of Plasticity,* Clarendon Press, Oxford, 1950.
3. Thomason, P.F. J. *Inst. Metals,* 1968, 96, p.360,
4. Brown, L.M. and Embury, J.D. *Proc. 3rd Int. Conf. on the Strength of Metals and Alloys,* Inst. of Metals, London, 1973, p.164.
5. Kudo, H. *Int. J. Mech. Sci.,* 1960, 2, p.102.
6. Alexander, J.M. *Quart, Appl. Math.* 1961, 19, p.32.
7. Thomason, P.F. *Acta Metall.,* 1981, 29, p.763.
8. Liu, C.T. and Gurland, J. *Trans. Am. Soc. Metals,* 1968, 61, p.156.
9. Hayden, H.W. and Floreen, S. *Acta Metall.,* 1969, 17, p.213.
10. Hill, R. *J. Mech. Phys. Solids,* 1957, 6, p.1.
11. Thomason, P.F. Acta Metall., 1985, 33, p.1079.
12. Thomason, P.F. *Acta Metall.,* 1985, 33, p.1087.
13. Rice, J.R. and Tracey, D.M. *J. Mech. Phys. Solids.,* 1969, 17, p.201.
14. Thomason, P.F. *J Appl. Mech.,* 1978, 45, p.678.
15. Thomason, P.F. *Acta Metall.,* 1982, 30, p.279.
16. Berg, C.A. *Inelastic Behaviour of Solids*, edited by M.F. Kanninen, p.171, McGraw-Hill, New York, 1970.
17. Gurson, A.L.   *J. Engng. Mater. Tech.,* 1977, 99, p.2.
18. Rudnicki, J.W. and Rice, J.R.   *J. Mech. Phys. Solids,* 1975, 23, p.371.
19. Yamamoto, H. *Int. J. Fract.,* 1978, 14, p.347.
20. Tvergaard, V. *Int. J. Fract.,* 1981, 178, p.389.
21. Prager, W. *Introduction to the Mechanics of Continua*, Ginn, Boston, 1961.
22. Thomason, P.F. *Prospects of Fracture Mechanics*, edited by G.C. Sih et al., p.3, Noordhoff, Netherlands, 1974.
23. Rice, J.R. *Theoretical and Applied Mechanics,* edited by W.T. Koiter, p.207, North Holland Publishing Co., Amsterdam, 1977.
24. Needleman, A. and Rice, J.R. *Mechanics of Sheet Metal Forming*, edited by D.P. Koistinen and N.-M. Wang, p.237, Plenum Press, New York, 1978.
25.   Mear, M.E. and Hutchinson, J.W.   *Mech. of Materials,* 1985, 4, p.395.
26. Tvergaard, V.   *J. Mech. Phys. Solids,* 1987, 35, p.43.
27. Hecker, S.S. *Constitutive Equations in Viscoplasticity: Computational and Engineering Aspects,* A.M.D. Vol. 20, edited by J.A. Stricklin and K.J. Saczalski, p.1, Am. Soc. Mech. Engrs., New York, 1976.
28. Lin, T.H. *Advanced in Applied Mechanics*, Vol. 11, edited by Chia-Shun Yih, p.256, Academic Press, New York, 1971.
29. Hecker, S.S.   *Metallurgical Trans.,* 1973, 4, p.985.

CHAPTER 4

The General Effects of Fluid Hydrostatic Pressure, Temperature
and Strain-Rate on the Mechanics of Ductile Fracture, and the
Conditions for Spontaneous Nucleation and Coalescence of Microvoids

4.1 The Effect of a Fluid Hydrostatic Pressure on the Nucleation and Growth of Microvoids

When a metal undergoes plastic deformation under the influence of a superimposed fluid hydrostatic pressure P, the effective mean-normal stress is given by the relation:

$$\sigma_{meff} = \sigma_m + P, \tag{4.1}$$

where σ_m is the mean-normal stress for plastic flow under zero fluid hydrostatic pressure and compressive values of P are taken to be algebraically negative. The effect of a fluid hydrostatic pressure on the nucleation of microvoids can therefore be estimated by replacing the mean-normal stress component with σ_{meff} in the particle-decohesion models of microvoid nucleation. The dislocation model of Brown and Stobbs [1] for particles of radius $r \leq 1\mu m$, equation (2.5), can thus be rewritten to account for a fluid hydrostatic pressure in the form:

$$\epsilon_1^n = Kr(\sigma_c - \sigma_m - P)^2 ; \tag{4.2}$$

where the parameters are defined in Chapter 2.1. A similar modification to the continuum model of Argon et al. [2], for particles of radius $r > 1\mu m$, equation (2.8), gives the following critical condition for decohesion of the particle/matrix interface:

$$\sigma_c = \bar{\sigma} + \sigma_m + P, \tag{4.3}$$

where $\bar{\sigma}$ is the equivalent stress. It is clear from the form of equations (4.2) and (4.3) that a highly compressive fluid pressure P can greatly increase the microvoid nucleation strain ϵ_1^n, and possibly even eliminate the microvoid nucleation

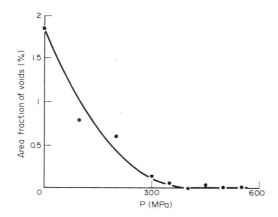

Fig. 4.1. Variation of the area-fraction of microvoids, adjacent to the ductile-fracture surfaces of α-brass tension specimens, with increasing fluid hydrostatic pressure P. (After I.E. French and P.F. Weinrich [4]).

mechanism if P is sufficiently high. For example, equation (4.2) implies that microvoid nucleation by particle decohesion cannot occur when $\sigma_m + P < -\sigma_c$ or $P < -(\sigma_c + \sigma_m)$; similarly equation (4.3) implies that microvoids will not be formed when $P < -(\bar{\sigma} + \sigma_m)$. This of course does not rule out the possibility of an alternative critical deviatoric-stress mechanism of particle damage intervening at highly negative mean-normal stresses, as pointed out in Chapter 2.2.

Experimental results for tensile tests under superimposed fluid hydrostatic pressure [3-7] confirm that microvoid nucleation is suppressed by the effects of a large hydrostatic pressure P and is virtually eliminated prior to fracture when P reaches a magnitude of the order of the uniaxial yield stress Y; this result is in broad agreement with the conditions given above for the suppression of the debonding mechanism of microvoid nucleation. A typical set of experimental results for the effects of a fluid hydrostatic pressure on microvoid nucleation in uniaxial tension specimens is given in Fig. 4.1 from the results of French and Weinrich [4] for an α-brass; these results show that as P is raised from zero to a value approaching 600 MPa the area fraction of microvoids adjacent to the ductile-fracture surfaces is reduced from ~ 2% to virtually zero. Clearly, at sufficiently high levels of fluid hydrostatic pressure P, ductile fracture can be suppressed merely by the effect of eliminating or suppressing the various mechanisms of microvoid nucleation (cf. Chapter 2.1), which will all be influenced to varying extents by a particular large hydrostatic-pressure. However, at more moderate hydrostatic pressures, where microvoid nucleation mechanisms can still operate, the subsequent growth of

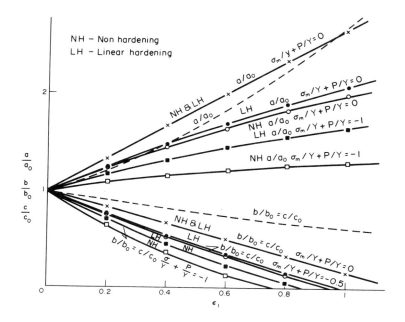

Fig. 4.2. The effect of increasing plastic strain ϵ_1 on the principal radii (a,b,c) of an initially spherical void ($a_o = b_o = c_o$) in uniaxial tension ($\nu = +1$) at various combinations of mean-normal stress $\sigma_m/$ Y and fluid hydrostatic pressure P/Y, for both non-hardening and linear-hardening material. The dotted lines show the changes in void dimensions that would occur without any amplification of the basic uniform strain field (ϵ_1, ϵ_2, ϵ_3).

microvoids will also be strongly influenced by the magnitude of the fluid hydrostatic pressure P.

An estimate of the effects of a superimposed hydrostatic pressure P on the growth of microvoids in a plastically deforming solid can be obtained from the Rice-Tracey model of void growth [8] (cf. Chapter 2.3). A simple modification to the void growth-rate equation (2.13), to allow for the effects of a superimposed hydrostatic pressure P, consists of replacing σ_m by σ_{meff} from equation (4.1). On integrating the resulting equation, by the method outlined in Chapter 2.3, for the case of an initially spherical void in a uniaxial-tensile flow field we obtain the results shown in Fig. 4.2 for the changing void geometry (a/a_o, b/b_o, c/c_o) with increasing tensile strain ϵ_1; the results are given for various effective mean normal stress levels ($\sigma_m + P$)/Y in both non-hardening and linear-hardening materials. Although the Rice-Tracey equations were not strictly intended for evaluation under negative mean-normal stresses, the results in Fig. 4.2 nevertheless give a sensible indication of the probable general effects of fluid hydrostatic pressure on the growth of microvoids. The first point to note from Fig. 4.2 is that for an effective mean-normal stress of zero (i.e., $P = -\sigma_m$) the extensional void growth a/a_0 is considerably lower and the transverse compressive growth $b/b_0 = c/c_0$ is considerably higher than the corresponding results in Fig. 2.11 for pure uniaxial tension with zero fluid hydrostatic pressure (i.e., $\sigma_m/Y = 0.333$). The results also show (Fig. 4.2) that changing the effective mean-normal stress level $(\sigma_m + P)$ to $-0.5Y$ and then to $-Y$ strongly accentuates the reduction in extensional growth a/a_0 and the increase in transverse compressive growth $b/b_0 = c/c_0$, and these effects are more pronounced for non-hardening plasticity in comparison to linear-hardening plasticity. In fact, the results suggest that complete transverse void closure $(b/b_0 = c/c_0 = 0)$ will occur at a strain $\epsilon_1 \approx 0.8$, when $(\sigma_m + P) = -Y$, and at $\epsilon_1 \approx 0.95$ when $(\sigma_m + P) = -0.5Y$. These results for transverse void closure cannot be regarded as exact, because the Rice-Tracey model is strictly beyond its limits of validity in this extreme void-growth regime. Nevertheless, the model applies in principle in this situation and the transverse void closure results in Fig. 4.2 give an approximate indication of the general effects of a large superimposed fluid hydrostatic pressure P on microvoid growth in ductile solids. It should also be remembered that in real materials the microvoids will usually contain an inclusion or second phase particle and complete transverse void closure will coincide with the elimination of a void at the sides of a particle; i.e., $c/c_0 = b/b_0 \to \sim 1$ rather than $c/c_0 = b/b_0 \to 0$.

Experimental confirmation of the general validity of the models described above, for the effects of a hydrostatic pressure P on the nucleation and growth of microvoids, can be obtained from the results of French and Weinrich [9] for the effects of hydrostatic pressure on the tensile deformation of a spheroidised 0.5% C steel. Their results for the variation in the number of microvoids N_v, at the centre of a necked tension specimen, with increasing true tensile strain $\epsilon_1 = \ln(A_0/A)$ are shown in Fig.4.3 for superimposed hydrostatic pressures P of zero (0.1 MPa), 300 MPa and 600 MPa, respectively. These results show that an increase in hydrostatic pressure P from zero to 300 MPa increases the microvoid nucleation strain ϵ_1^n from ~ 0.5 to ~ 0.9, in qualitative agreement with the decohesion models of microvoid nucleation [1,2], equations (4.2) and (4.3). A further 'increase' in hydrostatic pressure P to 600 MPa does not, however, increase the nucleation strain beyond $\epsilon_1^n \sim 0.9$, and this is in agreement with the suggestion in Chapter 2.2 (equation (2.11)) that at highly negative values of mean-normal stress σ_{meff} a decohesion mechanism of microvoid nucleation is likely to be replaced by a mechanism which operates at a critical deviatoric stress (Figs. 2.7(b) and 2.8). A further interesting feature of the results of French and Weinrich (Fig. 4.3) is that the initial gradients of the N_v/ϵ_1 curves show a reduction in magnitude as the hydrostatic pressure increases. This result is in qualitative agreement with the present void growth results in Fig. 4.2 which show an increasing suppression of the void-growth effect with increasing hydrostatic pressure P.

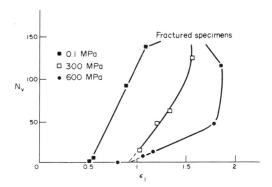

Fig. 4.3. Variation in the number of microvoids N_v, in the centre of a necked tension specimen, with increasing tensile strain ϵ_1 at various fluid hydrostatic pressures P. Results of I.E. French and P.F. Weinrich for a spheroidised 0.5% C steel [9].

4.2 The Effect of Hydrostatic Pressure on the Critical Conditions for Microvoid Coalescence.

In addition to the effects that a superimposed fluid-hydrostatic pressure can have in suppressing both the nucleation and growth of microvoids, it can also have the effect of suppressing the microvoid-coalescence phase of ductile fracture. The effect of fluid hydrostatic pressure P on the criterion of microvoid coalescence can be illustrated with the aid of the simple plane-strain model from Chapter 3.3. Replacing σ_m in equation (3.7) by σ_{meff} from equation (4.1) we obtain the following critical condition for incipient microvoid coalescence:

$$\frac{\sigma_n}{2k} \, (1 - \sqrt{V_f}) \; = \; \frac{1}{2} + \frac{\sigma_m}{2k} + \frac{P}{2k} \tag{4.3}$$

The influence of a large fluid pressure P in suppressing the microvoid coalescence effect is illustrated schematically in Fig. 4.4 for the critical condition (4.3). The schematic diagram shows that the hydrostatic pressure P effectively reduces the tensile magnitude of the flow stress $\sigma_1 + P = k + \sigma_m + P$, thus requiring an increased amount of plastic void-growth strain ϵ_1^G to satisfy the criterion (4.3) for incipient void coalescence (cf. Chapter 3.3 and 3.4). It follows from this that the generalised criterion for microvoid coalescence (Chapter 3.4) can be modified to show the effect of a fluid hydrostatic pressure P by replacing σ_1 by $\sigma_1 + P$ in equation (3.12) to give:

$$(\sigma_1^{1c} - \sigma_1 - P) \; \dot{\epsilon}_1^{1c} \; = \; 0 \, . \tag{4.4}$$

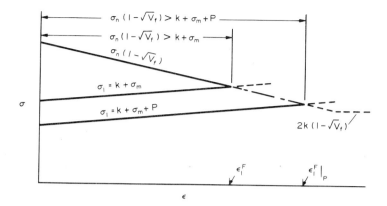

Fig. 4.4. A schematic diagram showing the critical conditions for incipient ductile fracture by microvoid coalescence, and the effect of a superimposed fluid hydrostatic pressure P in increasing the strain to ductile fracture.

Similarly, the generalised criterion for stable ductile flow, in the presence of microvoids, takes the form:

$$(\sigma_i^{1c} - \sigma_i - P)\,\dot\epsilon_i^{1c} > 0. \tag{4.5}$$

These equations show that the primary effect of a fluid hydrostatic pressure is to move the location of the active stress vector σ_i on the weak-dilational yield surface in the direction of a more negative mean-normal stress; this effect is illustrated in Fig. 4.5.

An interesting feature of the condition for microvoid coalescence, equations (4.3) to (4.5), is that a sufficiently large hydrostatic pressure should be capable of completely suppressing the microvoid coalescence mode of ductile fracture. For example, in the simple plane-strain model (equation (4.3)), σ_n cannot be less than 2k and thus the condition for microvoid coalescence will never be satisfied if the hydrostatic pressure P has a magnitude in excess of $(k - 2k\sqrt{V_f} - \sigma_m)$. This effect is illustrated schematically in Fig. 4.4 by the presence of the lower-limit $2k(1 - \sqrt{V_f})$ to the graph of $\sigma_n(1 - \sqrt{V_f})$. The virtually complete suppression of the void coalescence mode of ductile fracture at very high fluid hydrostatic pressures is confirmed by the work of Davidson and Ansell [10] where the fibrous (microvoid coalescence) mode of ductile fracture in a range of carbon steels (0.004% to 1.1% C) was eliminated for fluid hydrostatic pressures P in excess of 12 ~ 16 K bars, Fig. 4.6. The corresponding uniaxial yield stress Y of the various steels can be estimated from Davidson and Ansell's results to be in the range 8 ~ 16 K bars, hence the ratio P/Y to suppress the

Dilational – plastic yield surfaces

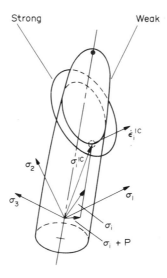

Fig. 4.5. The strong and weak dilational-yield surfaces for a solid containing microvoids and the effect of a fluid hydrostatic pressure P in
shifting the active stress vector σ_i away from the strong dilational surface; thus promoting ductility.

void coalescence mode of ductile fracture is of the order $P/Y = 1$. Now the theoretical ratio $P/2k$ for the suppression of void coalescence from the plane-strain model
can be writtten in the form:

$$\frac{P}{2k} = \frac{1}{2} - \sqrt{V_f} - \frac{\sigma_m}{2k}. \tag{4.6}$$

Using Bridgman's equations [11] to estimate $\sigma_m/2k$ at large plastic strains in a
deeply-necked tension specimen, we find that $\sigma_m/2k \sim 1$; hence for $V_f \sim 0.05$ we
obtain a critical pressure $P/2k \sim 0.75$ (or $P/Y \sim 0.87$), which is in good agreement
with Davidson and Ansell's result of $P/Y \sim 1$.

Estimates of the effect of a fluid hydrostatic pressure P on the tensile strain to ductile fracture in a body containing various volume fractions V_f of microvoids have
been made on the basis of the simple plane-strain model (equation (4.3)) [12] for
the case of both zero transverse stress (i.e., $\sigma_2 = 0$, $\sigma_m = k$) and varying mean-normal stresses σ_m estimated by Bridgman's equations [11]; the results are given in
Fig. 4.7(a) and (b), respectively. The theoretical results in Fig. 4.7(b) show a region

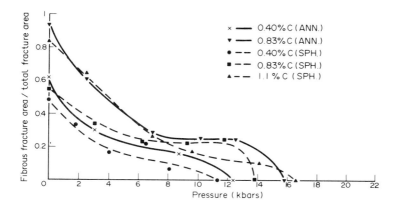

Fig. 4.6. Variation in percentage of fibrous ductile-fracture area to total fracture area, with increasing fluid hydrostatic pressure P, for various carbon steels. (After T.E. Davidson and G.S. Ansell, Trans. ASM, ASM International, 1968, Vol.61, p.242).

(P/2k < 0.5) where the tensile ductility increases in an approximately linear manner with increase in P/2k, in agreement with many experimental results [3-7,9,10]. The theoretical results also show a transition to extremely high ductility with further increase in hydrostatic pressure in the region P/2k ~ 0.75, the magnitude of the transition pressure tending to increase with volume fraction V_f, Fig. 4.7(b). The transition to virtually infinite ductility, with 100% reduction in area at the neck of a tension specimen at large fluid hydrostatic pressures, is well confirmed by experiment [6,10] and the results of Pugh and Green [6] for both an aluminium alloy and an electrolytically-refined high-conductivity copper are shown in Fig. 4.8; the similar forms of the theoretical and experimental results in Figs. 4.7(b) and 4.8, respectively, is clearly apparent. It should be noted, however, that for certain materials experimental results [3-5,7,9,10] do not show a transition to 100% ductile rupture at very high fluid pressures, due to a severe anisotropic mode of deformation that develops after approximately 90% reduction in area. This anisotropic-deformation effect leads to the development of approximately plane-strain deformation on crossed tangential-velocity discontinuities or shear bands [5]; consequently, this type of tension test is usually terminated by the intervention of intense shear deformation or slip along one of the velocity discontinuties or shear bands. These anisotropic shear-modes of rupture represent an anomalous effect in comparison to the void-coalescence modes of ductile fracture in approximately isotropic solids.

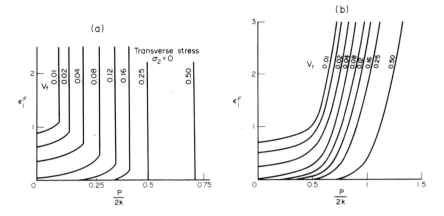

Fig. 4.7. Theoretical ductile-fracture strains ϵ_1^F at various fluid hydrostatic pressures P and microvoid volume fractions V_f for: (a) constant stress state $\sigma_2 = 0$, $\sigma_m = k$; (b) varying σ_2 and σ_m, similar to the variations in a uniaxial tension test with progressive external necking. (After P.F. Thomason, J. Inst. Metals, 1968, 96, p.360.)

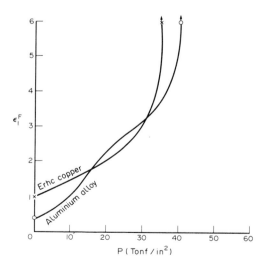

Fig. 4.8. Experimental results for the variation in fracture strain $\epsilon_1^F = \ln(A_o/A_F)$ with fluid hydrostatic pressure P; uniaxial tension tests on an aluminium alloy and an ERHC copper. (After H.Ll.D. Pugh and D. Green, Proc. I. Mech. Engrs., 1964-65, 179, p.417).

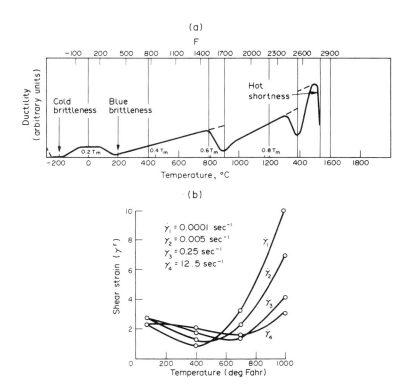

Fig. 4.9. The effect of temperature on the ductile-fracture strains of low-carbon steels at various strain-rates: (a) tension tests at a nominal strain-rate of 0.0017 sec^{-1}; (b) torsion tests at nominal shear strain-rates in the range 0.0001 to 12.5 sec^{-1}. ((a) After O.D. Sherby, Metals Engineering Quarterly, ASM International, 1962, Vol.2, p.3. (b) After C.F. Work and T.J. Dolan, Proc. ASTM, 1953, Vol.53, p.611, Copyright ASTM, reprinted with permission).

4.3 The Effects of Temperature and Strain-Rate on the Nucleation and Growth of Microvoids.

Experiments [13-16] show that the ductility of metals is strongly influenced by changes in both the temperature and strain-rate, and these effects are illustrated by the results in Fig. 4.9(a) for the ductility of a low-carbon steel at temperatures from 0 to 1 T_m [13], and in Fig. 4.9(b) for the ductility of a similar material at temperatures up to 0.5 T_m and strain-rates in the range 10^{-4} sec^{-1} to 12.5 sec^{-1} [14]. At high homologous temperatures, the ductile fracture process is strongly influenced by time-dependent effects such as grain boundary sliding and intergranular creep-controlled fracture; on the other hand, at low homologous temperatures, a mode of brittle 'cleavage' fracture intervenes when the ductile-brittle transition temperature is reached. The present work on ductile fracture is concerned only with true ductile-fracture processes in the range 0.1 to $\sim$ 0.7 T_m, at plastic strain-rates which are sufficiently high to virtually eliminate the effects of diffusional and creep-deformation mechanisms on the nucleation, growth and coalescence of microvoids [17]. In this temperature range, in the absence of dynamic strain-aging effects [14-18], temperature and strain-rate have an approximate inverse effect on material flow stress [14,16]; i.e., the effects on flow stress of an increase in strain-rate are broadly equivalent to a reduction in test temperature.

The models of microvoid nucleation [1,2] presented in Chapter 2.1 suggest that, in general, increased test temperatures and reduced strain-rates, which are likely to promote local recovery effects at particle sites, will reduce the maximum interface stress σ_T; thus making it more difficult to achieve the condition $\sigma_T = \sigma_c$ for decohesion of the particle/matrix interface. This is in broad agreement with experimental results [1,17,18] which show that microvoid nucleation is generally reduced by elevated temperatures and lower strain-rates, and can be virtually eliminated at high homologous temperatures [16] to give tensile failure by complete rupture (or the 100% reduction in area of an external neck). It follows from the inverse effects of temperature and strain-rate on the magnitude of the material flow stress, that increased strain-rates and reductions in test temperature should promote microvoid nucleation and this also is in broad agreement with experiment [1,17].

Once microvoids have been nucleated at the sites of particles and inclusions the subsequent growth of the voids is likely to be influenced by temperature and strain-rate primarily through their effect on the material yield stress and work-hardening rate. Restricting attention to the temperature range 0.1 to 0.7 T_m and strain-rate range 10^{-4} to 10^3 sec $^{-1}$, where inertia and stress-wave effects are negligible, it can be broadly assumed that increased test temperatures and reduced strain-rates will tend to reduce both the material yield stress and work-hardening rate. On this basis, we can use the quasi-static model of Rice and Tracey [8], and the results in Chapter 2.3, to estimate approximately the effects of temperature and strain-rate on void-growth rates using the non-hardening results (Figs. 2.11(a) and 2.12) to represent conditions of increased temperature or reduced strain-rate, and the linear-hardening results (Figs. 2.11(b) and 2.12) to represent conditions of reduced temperature or increased strain-rate. This interpretation of the results in Figs. 2.11 and 2.12 suggests that increased temperatures and reduced strain-rates will tend to promote the growth of microvoids, whilst the reverse effects will tend to reduce microvoid growth-rates. Unfortunately, there appears to be no definitive published experimental work on the effects of temperature and strain-rates on void-growth rates in the true ductile-fracture regime. However, an analysis of available experimental work [17,18] in this field shows no evidence to suggest that the above deductions are incorrect.

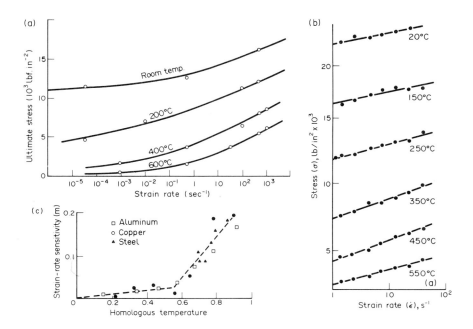

Fig. 4.10. The effect of temperature and strain-rate on: (a) the ultimate
tensile stress of commercially-pure aluminium; (b) the compressive flow stress at 40% compression for commercially-pure
aluminium; (c) the variation in the strain-rate sensitivity index
m with homologous temperature at 40% compression, for various materials. ((a) After A. Nadai and M.J. Manjoine, J.Appl.
Mech., 1941, Vol.8, p.77; copyright ASME, reprinted with permission. (b) and (c) After J.F. Alder and V.A. Phillips, J.Inst. Metals, 1954-55, 83, p.80).

4.4   The Effects of Temperature and Strain-Rate on the Critical Conditions for Microvoid Coalescence

Tension [16], torsion [14] and compression [19] tests on a range of materials at various temperatures and strain-rates show that, for a given test temperature, the
stress σ is related to the strain-rate $\dot{\epsilon}$ by a simple power law of the form:

$$\sigma = \sigma_o \dot{\epsilon}^m ; \qquad (4.7)$$

where σ_o is a stress related parameter and the strain-rate sensitivity index m is constant at a given temperature. Typical tension-test results for commercially pure
aluminium, at temperatures up to 600°C and strain-rates in the range 10^{-5} to 10^3 sec^{-1},

are shown in Fig. 4.10(a) from the work of Nadai and Manjoine [16]. These results show a similar strain-rate and temperature dependence to the 40% compression-test results of Alder and Phillips [19], at temperatures up to 550°C and strain-rates in the range 1 to 10^2 sec^{-1}, Fig. 4.10(b). Alder and Phillips [19] fitted a range of experimental results, on aluminium, copper and steel compression specimens, to equation (4.7) and obtained results for the strain-rate sensitivity index m at various homologous temperatures. These results are shown in Fig. 4.10(c) where it can be seen that strain-rate sensitivity of the flow stress is relatively small (m $\sim$ 0 to 0.025) for homologous temperatures up to $T/T_m \approx 0.5$; above $T/T_m = 0.5$, however, strain-rate sensitivity effects are more pronounced and m increases rapidly to m $\sim$ 0.2 at $T/T_m \approx 0.9$.

A departure from the normal inverse relationship between the effects of increasing test temperature and increasing test strain-rate on material flow stress, is found in the temperature range 200°C to 600°C for low carbon steels [16]. This is the so-called 'blue-brittle' range where dynamic strain-aging effects [18,20] have a strong influence on the material flow stess. Typical results for the effects of temperature and strain-rate on the material flow stress are given by the tension-tests of Nadai and Manjoine [16] in Fig. 4.11. These results show that the dynamic strain-aging effect in low-carbon steel can lead to negative values for the strain-rate sensitivity index m in the region of 400°C, Fig. 4.11; thus, giving a localised anomalous situation where an increase in strain-rate gives a reduction in flow stress.

It is clear from the results presented above that temperature and strain-rate effects can strongly influence the material yield stress and can therefore potentially have a strong influence on the conditions for microvoid coalescence. The primary source of this strong potential influence can be demonstrated with the aid of the simple plane-strain model of microvoid coalescence from Chapter 3.3. The plane-strain element with a rectangular array of voids is shown in Fig. 4.12 in the two possible response modes corresponding to: (a) weak dilational plasticity, where the complete element deforms homogeneously with strain-rate $\dot{\epsilon}_1 = V/L$; (b) strong dilational plasticity, where at incipient void coalescence along a single row of microvoids the mean strain-rate in the intervoid matrix is $\dot{\epsilon}_n \approx V/a$. It follows from equation (4.7) that the material yield-shear stress will be given approximately by $k \approx k_0 \dot{\epsilon}_1^m = k_0(V/L)^m$ in a homogeneous weak-dilational response, and by $k_n \approx k_0(\dot{\epsilon}_n)^m \approx k_0(V/a)^m$ at incipient void coalescence in a strong-dilational response. Hence, the ratio of the microscopic yield-shear stress k_n, in the intervoid matrix at incipient void coalescence, to the macroscopic yield-shear stress k is given by:

$$\left(\frac{k_n}{k}\right)_{SR} \approx \left(\frac{\dot{\epsilon}_n}{\dot{\epsilon}_1}\right)^m \approx \left(\frac{L}{a}\right)^m \qquad (4.8)$$

The strain-rate ratio $\dot{\epsilon}_n/\dot{\epsilon}_1 \approx L/a$ for typical alloy systems, at incipient microvoid coalescence, will be in the range $10^2 < L/a < 10^4$; in other words, the mean strain-

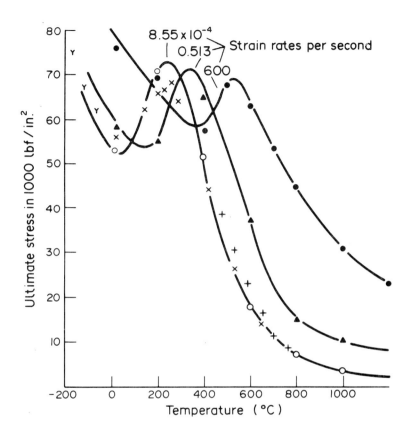

Fig. 4.11. The effect of temperature and strain rate on the ultimate tensile stress of a low-carbon steel. (After A. Nadai and M.J. Manjoine, J.Appl. Mech., 1941, Vol.8, p.77; copyright ASME, reprinted with permission).

rate in the intervoid matrix at the onset of microvoid coalescence can be up to four orders of magnitude greater than the homogeneous macroscopic strain-rate. The corresponding strain-rate sensitivity ratios of the yield shear stresses $(k_n/k)_{SR}$, at various values of the strain-rate sensitivity index m, are given in Fig. 4.12(c) and these results show that large yield-stress ratios of $(k_n/k)_{SR} > 2.5$ are likely to exist for quite small values (m $\approx$ 0.1) of the strain-rate sensitivity index.

The condition for incipient microvoid coalescence in the simple two-dimensional model, equation (3.7), can be readily modified to account for strain-rate sensitivity

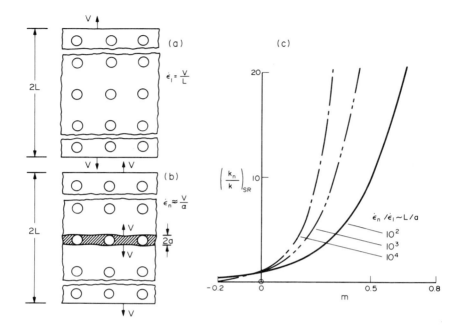

Fig. 4.12. The plane-strain model of ductile fracture showing: (a) the
mode of weak-dilational plasticity with a homogeneous strain-
rate $\dot{\epsilon}_1 = V/L$; (b) the mode of strong-dilational plasticity, at incipient void coalescence, where the localised intervoid strain-
rate $\dot{\epsilon}_n \approx V/a$; (c) the corresponding ratio of intervoid-matrix
yield stress to macroscopic yield stress $(k_n/k)_{SR}$, at incipient
void coalescence, for variations in the strain-rate sensitivity
index m.

of the material yield stress by rewriting it in the following form [21]:

$$\frac{\sigma_n}{2k_n} \; (1 - \sqrt{V_f}) \; \left(\frac{k_n}{k}\right)_{SR} \; = \; \frac{\sigma_1}{2k} \; = \; \frac{1}{2} + \frac{\sigma_m}{2k} \tag{4.9}$$

where the parameters are defined in Chapter 3.3. It should be noted at this stage
that there is little point in modifying this equation for the "law of mixtures" effect
of microvoids on the yield shear stress by equation (3.8), since $k_n/k = (1 - V_f)^{-1}$ will
in general have a magnitude very much closer to unity than the strain-rate sensitivity ratio $(k_n/k)_{SR}$, Fig. 4.12(c). It has been shown in Chapter 3.3, Fig. 3.5, that the constraint factor $\sigma_n/2k_n$ at incipient microvoid coalescence can never be less than unity
(i.e., $\sigma_n/2k_n \geq 1$); it follows, therefore, from the condition for microvoid coalescence, equation (4.9), that ductile fracture cannot occur when the strain-rate

sensitivity ratio satisfies the following condition:

$$\left(\frac{k_n}{k}\right)_{SR} > \frac{\left(\frac{1}{2} + \frac{\sigma_m}{2k}\right)}{(1 - \sqrt{V_f})} \ . \qquad (4.10)$$

When $\sigma_m/2k = 1$ and $V_f = 0.1$ the critical value of the strain-rate sensitivity ratio for the suppression of ductile fracture is $(k_n/k)_{SR} > 2.2$ and the results in Fig. 4.12(c) show that this could be achieved at quite small values of the strain-rate sensitivity index $m \sim 0.1$ to 0.2. This effect of strain-rate sensitivity on ductility is the probable explanation for the greatly enhanced tensile ductility of materials at sufficiently elevated temperatures and intermediate strain-rates, where external necking approaches 100% reduction in area [16,17]. The strain-rate sensitivity mechanism for the suppression of microvoid coalescence effects also undoubtedly contributes to the very high ductility observed in superplastic materials [17,21] where, in general, $m \geq 0.3$ [22,23]. However, for superplastic deformation in alloy systems it is also necessary to have a highly refined metallurgical structure and a strain-rate sensitivity sufficient to suppress plastic instability and localised external necking on the macroscopic scale [22,24,25].

It is clear from the form of the microvoid-coalescence criterion (4.9) and equation (4.8), that negative values of the strain-rate sensitivity index m will give $(k_n/k)_{SR} < 1$ and will thus tend to promote ductile fracture at smaller void-growth strains ϵ_1^G when the constraint factor $\sigma_n/2k_n$ is still large (i.e., $\sigma_n/2k_n \gg 1$). The reduction in ductility accompanying a negative strain-rate sensitivity index (i.e., $m < 0$) can be illustrated schematically [20] with the aid of a perfect-plasticity model displaying temperature and strain-rate sensitivity similar to the results of Nadai and Manjoine for a low-carbon steel [16], Fig. 4.11. The schematic yield stress/temperature/strain-rate relationships for the ideal material are shown in Fig. 4.13(a) for both the macroscopic level k and the intervoid matrix k_n, and on computing the corresponding strain-rate sensitivity ratios at various temperatures, $(k_n/k)_{SR}$ is found to be less than unity over the range $0.14 \leq T/T_o \leq 0.41$ and increase to a value $(k_n/k)_{SR} \geq 2$ at $T/T_o > 0.6$. The corresponding ductile-fracture strains ϵ_1^F for $\sigma_m/2k = 0.5$ and various volume fractions of voids, obtained [21] with the microvoid-coalescence criterion (4.9), are shown in Fig. 4.13(b) where $(k_n/k)_{SR} < 1$ corresponds to a low-ductility range and $(k_n/k)_{SR} \geq 2$ corresponds to a region of 100% ductile rupture or possibly superplastic flow. The low-ductility range, corresponding to a negative strain-rate sensitivity index m, is similar to the so-called "blue brittleness" range in low-carbon steels [13,16,18,20] and the effects of a negative strain-rate sensitivity index on the conditions for microvoid coalescence undoubtedly contributes to the observed reduction in ductility in this temperature range. There are, however, a number of additional metallurgical and mechanical features [18,20] that contribute to reduced ductility in the "blue-brittle" temperature range, including the highly unstable modes of plastic flow that can accompany the 'jerky' form of the plastic stress/strain curves in this temperature range.

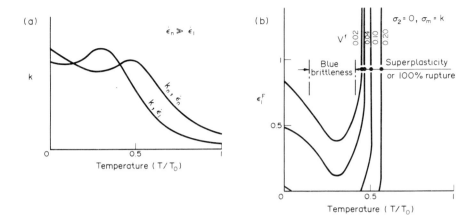

Fig. 4.13. (a) Schematic effect of temperature and strain rate on the yield
shear stress k of an ideal plastic material showing a temperature range of inverse strain-rate sensitivity (i.e., m < 0). (b) Corresponding theoretical ductile-fracture strains ϵ_1^F, for various
void volume fractions V_f and a constant stress state $\sigma_2 = 0$, σ_m
= k; showing temperature range of low ductility when m < 0
and transformation to extremely high ductility for sufficiently
large m. (After P.F. Thomason, Metal Science Journ., 1969, 3,
p.139).

The general effects of changes in the strain-rate-sensitivity index m from m > 0,
where material flow stress increases with strain rate, through m = 0, where flow
stress is insensitive to strain rate, to m < 0, where flow stress decreases with strain
rate, are illustrated schematically in Fig. 4.14. This figure shows the weak-dilational
yield surface for a given homogeneous macroscopic strain rate, with the active
stress vector σ_i, along with the strong dilational-yield surfaces corresponding to m
> 0, m = 0 and m < 0, respectively. When m > 0, the strong-dilational yield surface
is pushed away from the active stress vector σ_i on the weak-dilational surface by
the elevated yield-stress effect in the intervoid matrix, Fig. 4.14, hence an increased
tensile plastic strain ϵ_1 will be required to translate the strong-dilational surface
into contact with the active stress vector σ_i, thus giving increased ductility. On the
other hand, when m < 0 the strong-dilational surface is pushed towards the active
stress vector σ_i by the reduced yield stress in the intervoid matrix, Fig. 4.14, thus a
smaller tensile plastic strain is required to translate the strong-dilational surface
into contact with σ_i, giving reduced ductility. It should be emphasised at this point
that strain-rate sensitivity effects are readily incorporated into a plastic limit-load
model of ductile fracture [12,21,26], which exhibits a dual constitutive response
[26], because the primary effect of strain-rate sensitivity is felt predominantly in the

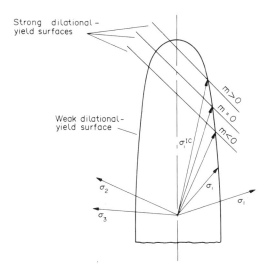

Fig. 4.14. The weak dilational-plastic yield surface for a given homogeneous strain rate, and the relative positions of the strong dilational-plastic yield surfaces for positive (m > 0), zero (m = 0)
and negative (m < 0) values of strain-rate sensitivity.

strong-dilational part of the dual-constitutive response (Chapter 3.3). On the other
hand, the Berg-Gurson model of ductile fracture (Chapter 3.4) is formulated only in
terms of second-order variations from a singular weak-dilational response, and the
effects of strain-rate sensitivity on the mechanics of ductile fracture are not readily
incorporated in the model.

4.5   Nucleation-Controlled Ductile Fracture by the Spontaneous Nucleation and Coalescence of Microvoids.

Metallographic studies of ductile fracture processes in alloy/particle systems with
a strong matrix/particle bond (cf. Chapters 1.6 and 2.1) [27-29] clearly indicate that
under certain conditions the normal sequential process of ductile fracture, by the
nucleation, growth and coalescence of microvoids, can be replaced by an apparently non-sequential process of spontaneous microvoid nucleation and coalescence (cf. Chapter 1.6). These effects can occur when there is a strong particle/mat-
rix bond because sustained tensile plastic flow can reduce the transverse interparticle spacing to such an extent that the plastic limit-load condition for the interparticle matrix is exceeded *before* the decohesion of the particle/matrix bond [30].

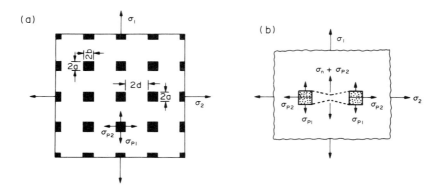

Fig. 4.15. The plane-strain model of ductile fracture with a regular array of rigid and strongly-bonded particles, showing the induced interface stresses σ_{P1} and σ_{P2} on the particle sides at incipient microvoid coalescence. (After P.F. Thomason, Metal Science Journ., 1971, 5, p.64).

Hence, when decohesion finally occurs the condition for incipient limit-load failure of the intervoid matrix is already satisfied and microvoid nucleation is followed immediately by microvoid coalescence. In this particular case, the void-growth process is no longer necessary to bring about ductile fracture.

The conditions under which this non-sequential mode of nucleation-controlled ductile fracture can occur in strongly-bonded alloy/particle systems can be developed from the simple plane-strain model of ductile fracture [12] (cf. Chapter 3.3) by assuming that square voids are filled with rigid and strongly-bonded particles [30], Fig. 4.15. The primary modification to the plane-strain model results from the presence of longitudinal σ_{P1} and transverse σ_{P2} mean-stresses generated at the rigid particle surfaces, and this leads to the following condition [30] for incipient limit-load failure of the interparticle matrix:

$$\left(\frac{\sigma_n + \sigma_{P2}}{2k}\right)(1 - \sqrt{V_f}) + \frac{\sigma_{P1}}{2k}\sqrt{V_f} = \frac{\sigma_1}{2k} = \frac{1}{2} + \frac{\sigma_m}{2k}; \qquad (4.11)$$

this reduces to the form of equation (3.7) when $\sigma_{P1} = \sigma_{P2} = 0$, i.e., when the rigid particles are replaced by voids. In general, the longitudinal stress σ_{P1} on the particle/matrix interface will have a magnitude $\sigma_{P1} \sim \sigma_1$ and equation (4.11) can therefore be rewritten [30]:

$$\frac{\sigma_n + \sigma_{P2}}{2k} = \frac{\sigma_1}{2k} = \frac{1}{2} + \frac{\sigma_m}{2k}; \qquad (4.12)$$

where it is to be emphasised that $(\sigma_n + \sigma_{P2})/2k$ is the constraint factor at incipient limit-load failure of the *interparticle* matrix and $\sigma_n/2k$ is the corresponding parameter for the equivalent *intervoid* matrix.

When the condition for incipient limit-load failure of the interparticle matrix (equation (4.12)) is satisfied, any tendency for internal necking to develop between adjacent particles will elevate σ_{P2} and thus further localised necking will be prevented [30]. The interparticle matrix can thus remain in a state of incipient limit-load failure, with continuing plastic flow, providing that the matrix/particle bond strength σ_c is greater than σ_{P2}. It follows from equation (4.12) that the condition for spontaneous microvoid nucleation and coalescence at strongly-bonded rigid particles is given by:

$$\frac{\sigma_{P2}}{2k} = \frac{1}{2} + \frac{\sigma_m}{2k} - \frac{\sigma_n}{2k} \geq \frac{\sigma_c}{2k}. \tag{4.13}$$

It should be noted that with sustained plastic flow, work-hardening effects will elevate the yield shear stress k, geometrical effects such as notches and external plastic necking will elevate σ_m, transverse contractional strains will reduce $\sigma_n/2k$ and σ_c will remain approximately constant. The effects of increasing plastic strain can therefore eventually lead to conditions satisfying the inequality (4.13) for the spontaneous nucleation and coalescence of microvoids. Since the plastic constraint factor $\sigma_n/2k_n$, for incipient limit-load failure of the intervoid matrix, approaches a lower limit of unity with increasingly large plastic strain ϵ_1 (cf. Chapter 3.3 and Fig. 3.5), the inequality (4.13) can be rewritten approximately for this case in the form:

$$\frac{\sigma_{P2}}{2k} \approx \frac{\sigma_2}{2k} \approx \frac{\sigma_m}{2k} - \frac{1}{2} \geq \frac{\sigma_c}{2k}. \tag{4.14}$$

This suggests that when plastic flow in a solid containing strongly-bonded hard particles has exceeded the point of incipient limit-load failure of the interparticle matrix, spontaneous microvoid nucleation and coalescence can intervene as soon as the transverse stress $\sigma_2 \geq \sigma_c$ or the mean normal stress $\sigma_m \geq \sigma_c + k$. These estimates of the critical values of σ_2 and σ_m, for spontaneous microvoid nucleation and coalescence, do not take into account the local elevation of the interface stress σ_{P2} above the macroscopic stress level σ_2 which can occur at small particles ($r < 1\ \mu m$) by dislocation mechanisms [1] and at large particles by stress concentration effects [2] (cf. Chapter 2.1). These critical values of σ_2 and σ_m are therefore "upper-bound" estimates of the conditions for spontaneous microvoid nucleation and coalescence. The associated problem of spontaneous microvoid-coalescence at the microvoid-nucleation strain ϵ_1^n, when the void-growth strain $\epsilon_1^G = 0$, is considered in Chapter 5.2.

REFERENCES

1. Brown, L.M. and Stobbs, W.M. *Phil. Mag.,* 1976, 34, p.351.
2. Argon, A.S., Im, J. and Safoglu, R., *Metallurgical Trans.* 1975, 6A, p.825.
3. French, I.E., Weinrich, P.F. and Weaver, C.W., *Acta Metall,* 1973, 21, p.1045.
4. French, I.E. and Weinrich, P.F., *Acta Metall,* 1973, 21, p.1533.
5. French, I.E. and Weinrich, P.F., *Metallurgical Trans,* 1975, 6A, p.785.
6. Pugh, H.Ll.D. and Green, D. *Proc. I. Mech. Engrs.,* 1964-65, 179, p.417.
7. Yajima, M. Ishii, M. and Kobayashi, M. *Int. J. Fract. Mech.,* 1970, 6, p.139.
8. Rice, J.R. and Tracey, D.M. *J. Mech. Phys. Solids,* 1969, 17, p.201.
9. French, I.E. and Weinrich, P.F. *Scripta Metall,* 1974, 8, p.87.
10. Davidson, T.E. and Ansell, G.S. *Trans. ASM,* 1968, 61, p.242.
11. Bridgman, P.W. *Studies in Large Plastic Flow and Fracture,* McGraw-Hill, N.Y., 1952.
12. Thomason, P.F. *J. Inst. Metals,* 1968, 96, p.360.
13. Sherby, O.D. *Metals Eng. Quart.,* 1962, 2, p.3.
14. Work, C.E. and Dolan, T.J. *Proc. ASTM,* 1953, 53, p.611.
15. Dieter, G.E. *Ductility,* Chapter 1, ASM Seminar 1967, p.1, American Society for Metals, Ohio, 1968.
16. Nadai, A. and Manjoine, M.J. *J. Appl. Mechanics,* 1941, 8, p.77.
17. Ashby, M.F., Gandhi, C. and Taplin, D.M.R. *Acta Metall,* 1979, 27, p.699.
18. Brindley, B.J. *Acta Metall,* 1968, 16, p.587.
19. Alder, J.F. and Phillips, V.A. *J. Inst. Metals,* 1954-55, 83, p.80.
20. Brindley, B.J. and Barnby, J.T. *Acta Metall,* 1968, 16, p.41.
21. Thomason, P.F. *Metal Science Journal,* 1969, 3, p.139.
22. Backofen, W.A., Azzarts, F.J., Murty, G.S. and Zehr, S.W. *Ductility,* Chapter 10, ASM Seminar 1967, p.279.
23. Chaudhari, P. *Acta Metall,* 1967, 15, p.1777.
24. Agrawal, S.P. *Superplastic Forming,* American Society for Metals, Ohio, 1985.
25. Mohamed, F.A., Ahmed, M.M.I. and Longdon, T.G. *Metallurgical Trans,* 1977, 8A, p.933.
26. Thomason, P.F. *Acta Metall,* 1981, 29, p.763.
27. Wilsdorf, H.G.F. *Mat. Sci. and Eng.,* 1983, 59, p.1.
28. Neumann, P. *Mat. Sci. and Eng.,* 1976, 25, p.217.
29. Le Roy, G., Embury, J.D., Edwards, G. and Ashby, M.F. *Acta Metall,* 1981, 29, p.1509.
30. Thomason, P.F. *Metal Science Journal,* 1971, 5, p.64.

CHAPTER 5

Theoretical and Empirical Models for
Estimating the Ductility of Metals

5.1 A Comparison of Microvoid Geometry in Two-Dimensional and Three-Dimensional Models of Ductile Fracture

Before developing numerical methods for estimating the magnitude of the critical void-growth strain ϵ_1^G in the total fracture-strain equation $\epsilon_1^F = \epsilon_1^n + \epsilon_1^G$, for a microvoid population N_i (cf. Chapter 3.1), it is useful to consider the disparity between the microvoid geometries for nominally equivalent volume-fractions of voids V_f in the two-dimensional and three-dimensional models. The unit cells containing the two-dimensional prismatic-elliptical void and three-dimensional ellipsoidal void, of similar (X_1, X_3) planar geometries, are shown in Figs. 5.1(a) and 5.1(b), respectively, for a state of tensile plastic strain $\epsilon_1 > 0$. Where it is to be noted that in the initial undeformed states ($\epsilon_1 = 0$) the two cubic unit-cells ($H_o \equiv L_o \equiv W_o$) would have contained, respectively a circular - cylindrical void ($a_o = c_o$) and a spheroidal void ($a_o = b_o = c_o$) in the two-dimensional and three-dimensional models. Now in the undeformed state the two-dimensional model has a void volume-fraction of $V_f = \frac{\pi}{4}$ $(\frac{c_o}{W_o})^2$ and the area fraction A_f of the void and matrix on the transverse (X_2, X_3) plane (Fig. 5.1a) is given by $A_f = c_o/W_o$; the corresponding expressions for the three-dimensional model are $V_f = \frac{\pi}{6}(\frac{c_o}{W_o})^3$ and $A_f = \frac{\pi}{4}(\frac{c_o}{W_o})^2$. The disparity between the initial geometry ratios c_o/W_o and transverse (X_2, X_3) area-fractions A_f, for a given volume fraction V_f, in the two- and three-dimensional models is illustrated by plotting the V_f and A_f expressions as shown Figs. 5.2(a) and (b), respectively. These results show that for a given value of c_o/W_o, which represents approximately the (X_1, X_3) planar mean void-diameter and spacing observed from a metallographic section, the volume fraction V_f and transverse area fraction A_f for the two-dimensional and three-dimensional models differ substantially. For example, with $c_o/W_o = 0.3$ the two-dimensional model gives $V_f = 0.071$ and $A_f = 0.3$ (Fig. 5.2a), and the three-

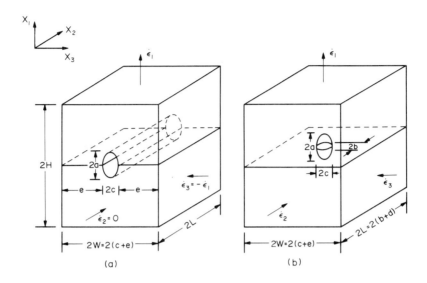

Fig. 5.1. The unit cells of similar (X_1, X_3) planar geometry for (a) two-dimensional models with prismatic elliptical voids and (b) three-dimensional models with ellipsoidal voids.

dimensional model gives $V_f = 0.014$ and $A_f = 0.073$ (Fig. 5.2b). Clearly these disparate geometrical effects in two-dimensional and three-dimensional unit cells, of nominally equivalent (X_1, X_3) planar ratio c_o/W_o, need to be taken into account in any attempt to develop a quantitative model of the ductile-fracture process. In fact, it will be shown in Section 5.4 that the results in Fig. 5.2 can be used to explain why two-dimensional and three-dimensional models of ductile fracture give nominally the same void-growth strains to fracture when compared on the basis of the initial microvoid volume-fractions V_f.

In formulating the critical conditions for incipient microvoid-coalescence in a plastic limit-load model of ductile fracture (cf. Chapter 3.3 and 3.4) it is necessary to develop appropriate expressions for the current area-fraction A_n of the intervoid matrix on the transverse (X_2, X_3) planes of the unit cells (Fig. 5.1), which is related to the current void area-fraction A_f by the expression $A_n = 1 - A_f$. From the geometry of the two-dimensional unit cell in a plane-strain ($\dot{\epsilon}_2 = 0$) flow field (Fig. 5.1a) it is clear that:-

$$A_n\big|_{2D} = 1 - \left(\frac{c}{w}\right) \equiv 1 - \left(\frac{c}{c_o} \cdot \frac{c_o}{W_o} \cdot \frac{W_o}{w}\right),$$

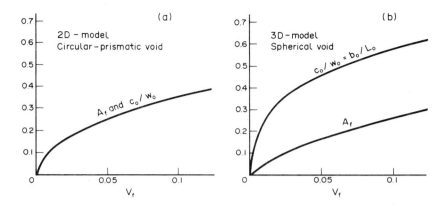

Fig. 5.2. The relationships between the transverse geometry ratio c_o/W_o of the unit cells and the transverse area fraction A_f for various void volume-fractions V_f in (a) two-dimensional and (b) three-dimensional models.

and on substituting $c_o/W_o = \sqrt{(4V_f/\pi)}$ and $W_o/W = \exp(\epsilon_1)$ we obtain the following expression for the transverse $(X_2, X_3$ plane) area-fraction of the intervoid matrix at the current tensile plastic strain ϵ_1 [1,2]:-

$$A_n\big|_{2D} = 1 - \sqrt{\frac{4 V_f}{\pi}}\left(\frac{c}{c_o}\right)e^{\epsilon_1} ; \qquad (5.1)$$

where c/c_o is given by R_3/R_o in the Rice-Tracey equation (2.16c) of Chapter 2.3. The corresponding expression for A_n in a three-dimensional tensile flow field ($\dot{\epsilon}_1 > 0$, $\dot{\epsilon}_2$, $\dot{\epsilon}_3$) can be obtained from the unit-cell geometry in Fig. 5.1(b) in the form [3]:-

$$A_n\big|_{3D} = 1 - \left(\frac{3\sqrt{\pi}}{4}V_f\right)^{2/3}\left(\frac{b}{b_o}\right)\left(\frac{c}{c_o}\right)e^{\epsilon_1} ; \qquad (5.2)$$

where $b/b_o \equiv R_2/R_o$ and $c/c_o \equiv R_3/R_o$ are given by equations (2.16b) and (2.16c), respectively. Equations (5.1) and (5.2), in conjunction with the appropriate values of the constraint factors for incipient limit-load failure of the intervoid matrix, now allow numerical estimates to be made of the critical loads to initiate microvoid coalescence by plastic limit-load failure of the intervoid matrix.

5.2 Theoretical Estimates of Ductile-Fracture Strains from the Two-Dimensional Plastic Limit-load Model of Ductile Fracture.

The critical condition for incipient microvoid-coalescence, given by equations (3.6) and (3.7) in Chapter 3.3, has a relatively simple form appropriate to a model with square-prismatic voids, where $A_n \approx (1 - \sqrt{V_f})$, which neglects the amplifying effect of mean-normal stress σ_m on void growth [1]. Replacing the $(1 - \sqrt{V_f})$ term by the expression (5.1) for a two-dimensional model with elliptical-prismatic voids gives the following critical condition for the termination of the void-growth strain ϵ_1^G at incipient ductile fracture, which now fully allows for the void-growth amplification effect of σ_m:-

$$\sigma_n \left(1 - \sqrt{\frac{4 V_f}{\pi}} \left(\frac{c}{c_o} \right) e^{\epsilon_1} \right) = \sigma_1 = k + \sigma_m \tag{5.3}$$

normalising this equation by the 'macroscopic' yield shear stress k and substituting the 'law of mixtures' relation (3.8) between k and the yield shear-stress of the intervoid matrix k_n gives:-

$$\frac{\sigma_n}{2k_n} \left(1 - V_f \right)^{-1} \left(1 - \sqrt{\frac{4V_f}{\pi}} \left(\frac{c}{c_o} \right) e^{\epsilon_1} \right) = \frac{\sigma_1}{2k} = \frac{1}{2} + \frac{\sigma_m}{2k} \tag{5.4}$$

where $\sigma_n/2k_n$ is the plastic constraint-factor for incipient limit-load failure of the intervoid matrix and σ_n is related to the *strong* dilational-plastic 'yield' stress σ_1^{1c} by the expression:-

$$\frac{\sigma_1^{1c}}{2k} = \frac{\sigma_1^{1c}}{2k_n} \left(1 - V_f \right)^{-1} = \frac{\sigma_n}{2k_n} \left(1 - V_f \right)^{-1} A_n . \tag{5.5}$$

It was shown in Chapter 3.4 that the limit-load constraint factor $\sigma_1^{1c}/2k_n$ can be regarded, to a close approximation, as a single valued function of the neck-geometry parameter $N = \frac{a}{w-c}$, and an expression of the form:-

$$\frac{\sigma_1^{1c}}{2k_n} = \frac{0.3}{N} + 0.6 \tag{5.6}$$

gives a close fit to the corresponding results in Fig. 3.14. Now the neck-geometry parameter N can be rewritten as a function of V_f and the current plastic strain ϵ_1 in the form:-

$$N = \frac{a}{c} \left(\frac{1 - A_n}{A_n} \right) , \tag{5.7}$$

where A_n is given by equation (5.1). Typical results for the variations in N with ϵ_1 for various V_f and σ_m values in non-hardening and linear-hardening materials were obtained in Chapter 3.4 and plotted in Fig. 3.11.

The critical condition for microvoid coalescence by plastic limit-load failure of the

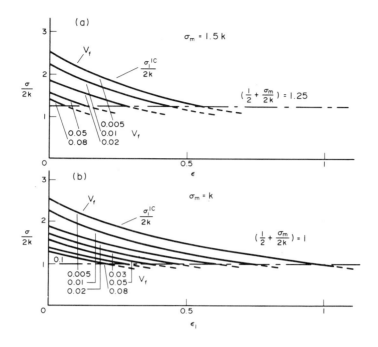

Fig. 5.3. Graphs of the left-hand ($\sigma_1^{1c}/2k$) and right-hand ($\frac{1}{2} + \sigma_m/2k$) sides of equation (5.8) showing the critical void-growth strains at incipient limit-load failure in the two-dimensional model for various void volume-fractions V_f at high (a) and low (b) mean-normal stresses.

intervoid matrix is therefore obtained from equations (5.4) to (5.7) in the form:-

$$\left(\frac{0.3 A_n}{a/c(1 - A_n)} + 0.6\right)\left(1 - V_f\right)^{-1} = \frac{1}{2} + \frac{\sigma_m}{2k}. \qquad (5.8)$$

Under normal conditions of ductile plastic void-growth the left-hand side of equation (5.8), which is equal to $\sigma_1^{1c}/2k$, exceeds ($\frac{1}{2} + \frac{\sigma_m}{2k}$) and microvoid coalescence is prevented (cf. Chapter 3.3 and 3.4), ductile fracture occurs only when the plastic void-growth strain is sufficient for the equality (5.8) to be satisfied. These effects were illustrated schematically in Fig. 3.6, where it was assumed for simplicity of presentation that the microvoid-nucleation strain was zero; thus $\epsilon_1^n = 0$ and $\epsilon_1^F = \epsilon_1^G$ from equation (3.1). In the following numerical estimates of ductile-fracture strains from the plane-strain model (equation (5.8)) it will be assumed that a certain unspecified microvoid-nucleation strain ϵ_1^n has already occured and thus the numerical results for the critical void-growth strains $\epsilon_1^G = \epsilon_1^F - \epsilon_1^n$ can be readily used to estimate the total fracture strain ϵ_1^F, whenever the corresponding microvoid-nucleation strain ϵ_1^n is known.

The void-growth strain ϵ_1^G to ductile fracture can be evaluated numerically from equations (5.1) and (5.8) by plotting the left-hand ($\sigma_1^{1c}/2k$) and right-hand ($\frac{1}{2} + \frac{\sigma_m}{2k}$) sides of equation (5.8) against ϵ_1; the critical condition for ductile fracture by plastic limit-load failure of the intervoid matrix is then satisfied when the graphs of $\sigma_1^{1c}/2k$ and ($\frac{1}{2} + \frac{\sigma_m}{2k}$) intersect. Numerical results for $\sigma_1^{1c}/2k$, with volume fractions of microvoids in the range $0.005 \leq V_f \leq 0.1$, are shown in Fig. 5.3(a) and (b) for constant mean normal stress levels of $\sigma_m = 1.5k$ and $\sigma_m = k$, respectively. The corresponding results for the variations in the critical void-growth strain ϵ_1^G, with void volume-fraction V_f, are shown in Fig. 5.4 for constant mean-normal stresses in the range $k \leq \sigma_m \leq 3k$. The results in Fig. 5.4 represent a two-dimensional plane-strain model of ductile fracture, with elliptical-prismatic voids, which takes into account both the amplifying effects of σ_m on void-growth and the disparity between the microscopic k_n and macroscopic k yield shear-stresses using a simple 'law-of-mixtures' model (cf. equation (3.8)). The numerical results therefore differ slightly from previously published results [1] for a model with rectangular prismatic voids, where it was assumed that $k_n \approx k$ and the void-growth amplification effect of σ_m was neglected. Nevertheless, the two sets of results show the same general trend with the critical void-growth strains ϵ_1^G rapidly decreasing with both increasing V_f and σ_m; again it is clear that the plastic limit-load model gives ductile-fracture strains of a realistic order which are very much lower than the physically unrealistic results from the Berg-Gurson model of ductile fracture (cf. Chapter 3.5 and Fig. 3.20).

An interesting feature of the criterion (5.8) for microvoid coalescence is that it allows an estimate to be made of the conditions for zero critical void-growth strain, where ductile fracture will occur spontaneously at the microvoid-nucleation strain; ie, $\epsilon_1^G = 0$ and $\epsilon_1^F = \epsilon_1^n$ (cf. Chapter 4.5). On setting $\epsilon_1 = 0$ in equation (5.1) and substituting the resulting expression for A_n into equation (5.8) we obtain:-

$$\frac{\sigma_1^{1c}}{2k} = \frac{\left(0.3 \left(\sqrt{\frac{\pi}{4V_f}} - 1 \right) + 0.6 \right)}{(1 - V_f)} = \frac{1}{2} + \frac{\sigma_m}{2k}. \qquad (5.9)$$

This condition for spontaneous ductile fracture at the microvoid-nucleation strain ϵ_1^n is plotted in Fig. 5.5, where the mean-normal stress level σ_m determines the microvoid volume fraction V_f above which ductile fracture is likely to occur with a zero critical void-growth strain (ie. $\epsilon_1^G = 0$). For example, the results suggest that when $\sigma_m = 2k$, the ductile-fracture strain $\epsilon_1^F = \epsilon_1^n$ for all void volume-fractions $V_f \geq 0.055$. There is no definitive experimental work to check the validity of these results on a quantitative basis, however, the phenomenon of spontaneous ductile fracture at the microvoid-nucleation strain is well documented under certain experimental conditions [12 - 14] (cf. Chapters 1.6 and 4.5).

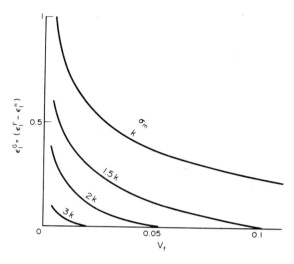

Fig. 5.4. Variation in void-growth strains to fracture for the two-dimensional plane-strain model, with various void volume-fractions V_f and mean-normal stresses σ_m.

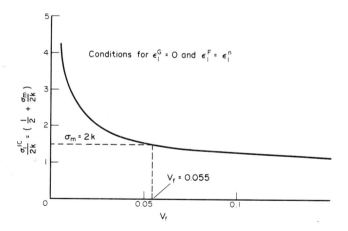

Fig. 5.5. The conditions for spontaneous microvoid coalescence at the microvoid-nucleation strain, giving the critical void volume-fraction V_f for any particular mean-normal stress σ_m.

5.3 Theoretical Estimates of Ductile-Fracture Strains from a Three-Dimensional Model of Ductile Fracture.

To some extent a two-dimensional model of ductile fracture, with elliptical-prismatic voids as shown in Fig. 5.1(a), represents an inadequate *quantitative* model of real ductile-fracture processes in metals where the microvoids generally have a form closer to the ellipsoidal geometry shown in Fig. 5.1(b). It is therefore desirable, in making quantitative estimates of ductile-fracture strains from the plastic limit-load model, to develop a three-dimensional model for incipient microvoid coalescence between ellipsoidal voids in unit cells of the form shown in Fig. 5.1(b). This approach not only leads to the possibility of obtaining more physically-appropriate results for ductile-fracture strains, to compare with experimental results, but also allows an assessment to be made of the general accuracy of results from two-dimensional models; this latter assessment is likely to be important under conditions where it is difficult to obtain an adequate model of a particular three-dimensional plastic limit-load condition.

The critical condition for incipient microvoid coalescence in a unit cell with an ellipsoidal void (Fig. 5.1b) is given by equation (3.13) and this can be rewritten in the form:-

$$\sigma_1^{1c} = \sigma_n A_n = \sigma_1 . \tag{5.10}$$

Substituting for A_n from equation (5.2) and rewriting σ_1 in terms of the mean-normal stress σ_m, 'macroscopic' uniaxial yield stress Y and Lode variable v we obtain an expression [3] which can be normalised by the 'law-of-mixtures' yield stress relation $Y = (1 - V_f) Y_n$ to give:-

$$\frac{\sigma_1^{1c}}{Y} = \left(\frac{\sigma_n}{Y_n}\right)(1 - V_f)^{-1}\left(1 - \left(\frac{3\sqrt{\pi}}{4} V_f^{2/3}\right)\left(\frac{b}{b_o}\right)\left(\frac{c}{c_o}\right)e^{\epsilon_1}\right) = \frac{\sigma_m}{Y} + \frac{3 + v}{3\sqrt{v^2 + 3}} ; \tag{5.11}$$

where σ_n/Y_n is the plastic constraint-factor for incipient limit-load failure of the intervoid matrix, Y_n is the 'microscopic' uniaxial yield stress of the intervoid matrix and σ_1^{1c} is the strong dilational-plastic 'yield' stress. Under conditions of ductile-plastic flow the left-hand side of equation (5.11) involving $\sigma_1^{1c}/Y = \sigma_n/Y_n (1 - V_f)^{-1} A_n$ exceeds the right-hand side involving $\sigma_m/Y + \frac{1}{3} (3 + v) / \sqrt{v^2 + 3} = \sigma_1/Y$ and ductile fracture is thus prevented; ductile fracture by microvoid coalescence only occurs when the equality (5.11) is just satisfied after a sufficient void-growth strain $\epsilon_1 = \epsilon_1^G$ (cf. Chapter 3.4 and Fig. 3.6).

In solving equation (5.11) it is necessary to obtain an expression for the plastic constraint factor σ_n/Y_n for incipient limit-load failure of the current intervoid matrix, in a unit cell with an ellipsoidal void (Fig. 5.1b), which is analogous to the two-dimensional expression in equation (5.6). Unfortunately the problem of finding an analytical

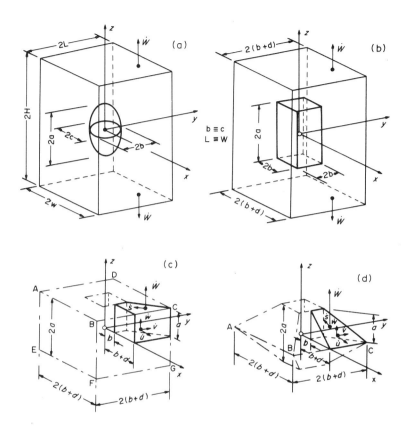

Fig. 5.6. The unit cell with (a) an ellipsoidal void and (b) the equivalent square-prismatic void; also showing the parallel (c) and triangular (d) velocity fields for the intervoid matrix. (After P.F. Thomason [4]).

solution for the plastic constraint-factor in a unit cell of incremental-plastic material, with an ellipsoidal void, appears to be intractable. However, a simple transformation of void shape into a geometrically equivalent square-prismatic void with the same principal dimensions as the ellipsoidal void (Fig. 5.6(a) and (b)), gives an intervoid-matrix boundary in each unit cell which is compatible with relatively simple kinematically-admissible velocity fields (Figs. 5.6(c) and (d)) of 'parallel' (Fig. 5.6c) and 'triangular' (Fig. 5.6d) forms. It is therefore possible to obtain good upper-bound solutions [4] for the plastic constraint-factor σ_n/Y_n, at incipient limit-load failure of the intervoid matrix of a unit cell, for a wide range of void and cell geometries [3,4]. It needs to be emphasised at this point that the use of an upper-bound constraint factor σ_n/Y_n in the microvoid-coalescence criterion (5.11) will always lead to an *overestimate* of the actual critical void-growth strain $\epsilon_1^G = \epsilon_1^F - \epsilon_1^n$ because, by the upper-bound theorem [5], $(\sigma_n/Y_n)_{UB}$ is always *greater* than the actual value of σ_n/Y_n.

The kinematically-admissible velocity fields $(\dot{u},\dot{v},\dot{w})$ at any point (x,y,z) in the generating segments of the parallel (Fig. 5.6c) and triangular (Fig. 5.6d) plastic zones have the following forms (4):-

(a) the parallel velocity field,

$$\dot{u} = \frac{\dot{w}}{2ax} \left[(b + d)^2 - x^2 \right] \tag{5.12a}$$

$$\dot{v} = \frac{\dot{w}y}{2ax^2} \left[(b + d)^2 - x^2 \right] \tag{5.12b}$$

$$\dot{w} = \frac{\dot{w}z}{a} \tag{5.12c}$$

$$\dot{s} = \sqrt{\dot{u}^2 + \dot{v}^2} \; ; \tag{5.12d}$$

(b) the triangular velocity field,

$$\dot{u} = \frac{\dot{w}d}{2ax} \left[(b + d) + x \right] \tag{5.13a}$$

$$\dot{v} = \frac{\dot{w}yd}{2ax^2} \left[(b + d) + x \right] \tag{5.13b}$$

$$\dot{w} = - \frac{\dot{w}zd}{2ax} \tag{5.13c}$$

$$\dot{s} = \frac{\dot{w}}{2x} \left[(b + d) + x \right] \left[1 + \left(\frac{d}{a}\right)^2 + \left(\frac{y}{x} \cdot \frac{d}{a}\right)^2 \right]^{\frac{1}{2}} \; ; \tag{5.13d}$$

where $\dot{s}$ represents the tangential velocity-discontinuities on the plastic-rigid boundaries of the plastic zones [4]. The rate of internal energy dissipation $\dot{I}$ in the limit-load plastic zone is given by [5-7]:-

$$\dot{I} = \sqrt{2/3} \, Y_n \int_V \left[\dot{\epsilon}_x^2 + \dot{\epsilon}_y^2 + \dot{\epsilon}_z^2 + \frac{1}{2} (\dot{\gamma}_{xy}^2 + \dot{\gamma}_{xz}^2 + \dot{\gamma}_{yz}^2) \right]^{\frac{1}{2}} dV + \frac{Y_n}{\sqrt{3}} \int_S \dot{s} \, dS \tag{5.14}$$

where V is the volume of the plastic zone, S is the surface area of the rigid/plastic

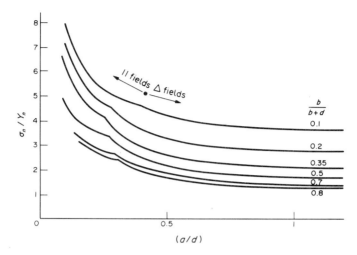

Fig. 5.7. The upper-bound constraint factors σ_n/Y_n for incipient plastic flow in the intervoid matrix of the three-dimensional unit cells, with parallel and triangular velocity fields (After P.F. Thomason [4]).

boundary and the strain-rate components are related to the velocity-field components by the equations:-

$$\dot{\epsilon}_x = \frac{\partial \dot{u}}{\partial x} ; \quad \dot{\epsilon}_y = \frac{\partial \dot{v}}{\partial y} ; \quad \dot{\epsilon}_z = \frac{\partial \dot{w}}{\partial z} ; \quad \dot{\gamma}_{xy} = \left(\frac{\partial \dot{u}}{\partial y} + \frac{\partial \dot{v}}{\partial x} \right) ;$$

$$\dot{\gamma}_{yz} = \left(\frac{\partial \dot{v}}{\partial z} + \frac{\partial \dot{w}}{\partial y} \right) ; \quad \dot{\gamma}_{xz} = \left(\frac{\partial \dot{u}}{\partial z} + \frac{\partial \dot{w}}{\partial x} \right) . \qquad (5.15)$$

Evaluating equations (5.14) and (5.15) with the appropriate velocity-field equations (5.12 or 5.13), gives an expression for $\dot{I}$ which can be equated to the rate of external energy dissipation $\dot{E} = \sigma_n A_n \dot{W}$ to give the upper-bound constraint factor σ_n/Y_n [4]. The numerical results for σ_n/Y_n, from the parallel and triangular velocity fields, are given in Fig. 5.7 in terms of the three-dimensional neck geometry parameters a/d and b/(b+d), respectively; the parallel velocity-fields giving the least upper-bounds for low values of a/d, the triangular velocity-fields giving the least upper-bounds at higher values of a/d.

In order to use equation (5.11) for making numerical estimates of the critical void-growth strains ϵ_f^G at incipient microvoid coalescence, it is necessary to obtain a closed-form empirical expression for the constraint factor σ_n/Y_n which gives close agreement with the upper-bound results in Fig. 5.7. It is readily shown [4] that the following empirical relation gives good agreement with the numerical results over the required ranges of the a/d and b/(b+d) neck-geometry parameters:-

$$\frac{\sigma_n}{Y_n} = \frac{0 \cdot 1}{(a/d)^2} + \frac{1 \cdot 2}{\left(\dfrac{b}{b+d}\right)^{\frac{1}{2}}} \cdot \tag{5.16}$$

This empirical expression for the plastic constraint factor at incipient limit-load failure of the intervoid matrix can therefore be substituted into equation (5.11) to give the critical condition for incipient microvoid coalescence and ductile fracture. For the case of uniaxial tension, where $\nu = +1$ and $b = c$, equation (5.11) then takes the form:-

$$\left(\frac{0 \cdot 1}{(a/d)^2} + \frac{1 \cdot 2}{\left(\dfrac{b}{b+d}\right)^{\frac{1}{2}}}\right)\left(1 - V_f\right)^{-1}\left(1 - \left(\frac{3\sqrt{\pi}\,V_f}{4}\right)^{\frac{2}{3}}\left(\frac{b}{b_0}\right)^2 e^{\epsilon_1}\right) = \frac{\sigma_m}{Y} + \frac{2}{3} \cdot \tag{5.17}$$

In the early stages of ductile void-growth the left-hand side ($\equiv \sigma_1^{1c}/Y$) of equation (5.17) exceeds the right-hand side ($\equiv \sigma_1/Y$) and microvoid coalescence is prevented (cf. Chapter 3.4). With continuing void-growth strain the equality (5.17) will eventually be satisfied and ductile fracture can then proceed by plastic limit-load failure of the intervoid matrix; the initial incipient-fracture surface tending to develop along a characteristic surface of the plastic velocity-field (cf. Chapter 3.2 and 3.3). It must be remembered, however, that once the first small area of macroscopic fracture-surface has developed in a continuous plastic velocity-field, the presence of the small ductile-fracture surface can profoundly change the nature of subsequent plastic flow. For example, the resulting stress-concentration effect and loss of plastic constraint can introduce tangential velocity-discontinuities into the plastic-flow field along which the subsequent ductile-fracture surfaces are likely to propagate (cf. Chapters 1.6, 6.5 and 8.2). It is necessary therefore always to regard the formation of the initial macroscopic fracture-surface and the subsequent propagation to form the complete ductile-fracture surface, as two quite distinct and separate processes. The formation of the small initial fracture-surface is the critical event governed by equation (5.17), and the subsequent propagation of this initial macroscopic fracture-surface will be strongly influenced by the development of any discontinuities in the plastic velocity-field. Of course, the formation of the *initial* macroscopic fracture surface *is* the important critical event in the mechanism of ductile fracture; once a small macroscopic fracture-surface has developed, the structural integrity of an engineering component is no longer guaranteed and the propagation of this crack into a complete ductile-fracture surface often becomes a problem of largely academic interest.

The critical void-growth strains ϵ_1^G to bring about incipient microvoid- coalescence and ductile fracture in pure uniaxial tension ($\nu = +1$), at a constant mean-normal stress level $\sigma_m/Y = 0.833$, have been evaluated from equation (5.17) and are shown

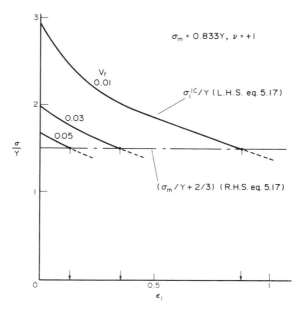

Fig. 5.8. Graphs of the left-hand (LHS) and right-hand (RHS) sides of equation (5.17) showing convergence at the critical strains for incipient limit-load failure in the three-dimensional model, with various void volume-fractions V_f; uniaxial tension ($\nu = +1$) at constant mean-normal stress σ_m.

in Fig. 5.8 for various void volume-fractions V_f. For a given value of σ_m/Y the right-hand side of equation (5.17) remains constant and the critical void-growth strain ϵ_1^G for a given V_f is reached when the left-hand side of (5.17) has been reduced sufficiently by continuous void-growth to bring about an equality, Fig. 5.8; ie increasing tensile strain ϵ_1 causes a continuous reduction in the σ_1^{1c}/Y (LHS) term, to bring about equality with the σ_1/Y (RHS) term at the critical void-growth strain ϵ_1^G. (NOTE: an approximate method of evaluating ϵ_1^G, under conditions where σ_m/Y increases with the void-growth strain, will be presented in Chapter 6.4).

The critical void-growth strains ϵ_1^G at incipient ductile-fracture were obtained by the above procedure for a wide range of σ_m/Y and V_f values in a pure tensile-flow field ($\nu = +1$) and the results are shown in Fig. 5.9. These results confirm that the plastic limit-load model of ductile fracture [1-4],for the three-dimensional case of approximately ellipsoidal voids, gives physically realistic estimates of the critical void-growth strains $\epsilon_1^G = (\epsilon_1^F - \epsilon_1^n)$ at incipient ductile-fracture, and it must be remembered that these results are in fact slight *overestimates* of the actual theoretical void-

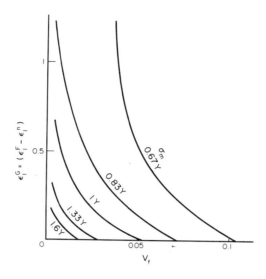

Fig. 5.9. Variation in void-growth strains to fracture ϵ_1^G for various void volume-fractions V_f and constant mean-normal stresses σ_m; uniaxial tension ($\nu = +1$) in the three-dimensional model.

growth strains due to the upper-bound nature of the present constraint factors σ_n/Y_n. In contrast to the realistic results obtained from the plastic limit-load model of ductile fracture, the equivalent uniaxial-tension results from the Berg-Gurson model of ductile fracture [8-11] (cf. Chapter 3.5) show no tendency to form localised deformation bands even at very high mean-normal stress levels (ie $\sigma_m/Y \geqslant 1.333$), because the h_{mcr} term in equation (3.25) remains strongly negative however large the uniaxial tensile strain. This leads to the absurd prediction, from the Berg-Gurson model, that ductile fracture will never occur in a state of uniaxial tension.

5.4 A Comparison of the Critical Void-Growth Strains from Two-Dimensional and Three-Dimensional Models of Ductile Fracture.

A comparison of the critical void-growth strains from the two-dimensional (Fig. 5.4) and three-dimensional (Fig. 5.8) models of ductile fracture show an interesting formal similarity when the two sets of results are plotted against void volume-fraction, for the various mean-normal stress levels. An explanation for the approximate numerical agreement between the two ductile-fracture models, when compared on the basis of the initial volume-fraction of voids V_f, can be obtained from the void- and matrix-geometry results in Fig. 5.2, along with the expressions for the

Table 5.1. A comparison of the parameters describing the geometry of the inter-void matrix and the corresponding constraint factors at limit-load failure, in two-dimensional (2D) and three-dimensional (3D) models of ductile fracture.

V_f	c_o/W_o		A_f		N	$\dfrac{b}{b+d}$	$\dfrac{a}{d}$	$\sigma_n/2k_n$	σ_n/Y_n	$\sigma_1^{1c}/2k$	σ_1^{1c}/Y
	2D	3D	2D	3D	2D	3D	3D	2D	3D	2D	3D
0.04	0.226	0.424	0.226	0.141	0.292	0.424	0.736	2.093	2.05	1.688	1.834
0.06	0.276	0.486	0.276	0.185	0.381	0.486	0.946	1.92	1.78	1.479	1.544
0.08	0.319	0.534	0.319	0.224	0.469	0.534	1.146	1.806	1.60	1.337	1.35
0.10	0.357	0.576	0.357	0.260	0.555	0.576	1.358	1.788	1.59	1.278	1.307

2D plastic constraint factor $\sigma_n/2k_n$ from equations (5.5) to (5.7) and the 3D plastic constraint factor σ_n/Y_n from equation (5.16). Using these results a further set of data can be constructed, as shown in Table 5.1, where for each initial void volume-fraction V_f the corresponding values of c_o/W_o and A_f for the 2D and 3D models are obtained from Fig. 5.2, and these are found to have a wide disparity. The geometry parameters N (2D) and $b/(b+d)$, a/d (3D) for the intervoid matrix can now be obtained from c_o/W_o and A_f, and the corresponding constraint factors estimated from equation (5.5) to (5.7) and (5.16). On converting the constraint factors to the limit-load form $\sigma_1^{1c}/2k$ for the two-dimensional model and σ_1^{1c}/Y for the three-dimensional model, it can be seen from Table 5.1 that these critical 'limit-load' stresses are in fact in very close agreement, for each value of V_f, even though the intervoid matrix geometries of the 2D and 3D models differ substantially. It is readily shown that these effects are preserved with increasing amounts of void-growth strain and this is clearly the explanation for the close agreement observed between the ductile-fracture strains obtained from two-dimensional and three-dimensional models of ductile fracture, when correlated on the basis of the void volume-fraction V_f. The general effects discussed above are responsible for the successful modelling of the ductile-fracture problem in terms of two-dimensional plane-strain models of void growth and coalescence, as shown in Chapter 3. In the subsequent work, however, the more physically appropriate three-dimensional model of ductile fracture will be used, wherever possible, to obtain numerical estimates of ductile-fracture strains, in a particular plastic flow-field, for comparison with the corresponding experimental results.

5.5   Empirical Methods of Estimating Ductile-Fracture Strains. The Phenomenological Model of Oyane.

In the absence of a reliable quantitative model for estimating the ductility of a material undergoing large plastic flow, in a metal-forming or metal-working process, a number of phenomenological models of metal-processing ductility have been developed but with only limited success [15,16,17]. Of these models the most successful is that of Oyane [17] which is based on the assumption that ductile fracture will occur at a certain critical volumetric-dilational strain $\epsilon = \frac{1}{3} \ln (V_c/V_o)$ (cf. Chapter 2.3 and equation (2.18)) as the overall density of a metal decreases with increasing plastic void growth. Unfortunately there are strong objections to the assumption that ductile fracture is controlled by the dilational strains, and these are based on the fact that in general ductile plastic flow is accompanied by large extensional-growth of the microvoids, in the direction of maximum principal strain, coupled with a relatively small transverse void-growth (cf. Chapter 2.3 and Figs. 2.11 and 2.12). Thus ductile fracture usually develops after relatively large "deviatoric" or shape-changing void growth and relatively small "dilational" or volume-changing growth; only at very large tensile mean-normal stresses will the deviatoric and dilational components of void growth approach the same order of magnitude. Consequently, there is not in general a direct one-to-one relationship between the mean-volumetric or dilational strain and the crucially important geometry of the intervoid matrix. For example, under pure torsional plastic flow the mean stress is virtually zero throughout the deformation path, thus giving virtually zero volumetric dilation of microvoids coupled with large extensional void growth. In this case ductile fracture in pure torsion is clearly not directly related to either the volumetric void-growth or the dilational strain ϵ. Nevertheless, the Oyane empirical criterion of ductility [17] is successful to the extent that it identifies some of the important parameters that determine ductility in metalworking processes. It can also be shown [18] that Oyane's empirical equation is consistent, to a first approximation, with the more correct equations for estimating ductile-fracture strains based on the plastic limit-load model of ductile fracture (Chapter 3), but only at large tensile mean-normal stress levels.

The Oyane model is based on a proposed theory of plasticity for porous materials [19] in which the dilational stress/strain relation is assumed to have the form:-

$$d\epsilon = \frac{d\bar{\epsilon}}{\Lambda} \left(\frac{\sigma_m}{\bar{\sigma}} + A_o \right),   \tag{5.18}$$

where $\bar{\sigma}$ and $d\bar{\epsilon}$ are the equivalent stress and strain increment, respectively, Λ is a function of the relative density of the porous solid and A_o is a material constant. When there are no voids present $\Lambda \to \infty$ and the proposed theory of plasticity reduces to the well known Levy-Mises form [19]. Oyane integrated equation (5.18) over the total ductile strain-path to give:-

$$\int_o^{\epsilon_f} \frac{\Lambda d\epsilon}{A_o} = \int_o^{\bar\epsilon_f} \left(1 + \frac{\sigma_m}{A_o\bar\sigma} \right) d\bar\epsilon \, , \tag{5.19}$$

and it was then assumed that the left-hand integral would represent a material constant B_o thus giving the following criterion of ductile fracture:-

$$\int_o^{\bar\epsilon_f} \left(1 + \frac{\sigma_m}{A_o\bar\sigma} \right) d\bar\epsilon = B_o \, . \tag{5.20}$$

With this form of semi-empirical equation the material constants A_o and B_o are determined by experiments on the appropriate material and it has been observed that fracture data from metalforming processes on a given material can often comply with this type of criterion to a reasonable degree of accuracy [17]. However, there are also many examples where additional "weighting factors" must be included in the integrand if reasonable agreement is to be obtained over a wide range of experimental results [17].

As pointed out above, the main limitation on the validity of the phenomenological models of ductile fracture, similar to equation (5.20), is that they are based on inadequate assumptions on the physical nature of the ductile-fracture process in metals, largely neglecting the mechanisms of microvoid nucleation and assuming that ductile fracture is governed by a critical volumetric or 'dilational' strain independently of the 'deviatoric' strain. The fact that ductile fracture readily occurs in pure torsion tests is, as previously stated, confirmation of the inadequacy of the theoretical basis for this type of empirical model. In the subsequent treatment of metalworking-ductility problems, in Chapter 7, all estimates of ductile-fracture strains will be made only in terms of the physically realistic plastic limit-load model of ductile fracture [1-4, 16].

REFERENCES

1. Thomason, P.F. *J. Inst. Metals,* 1968, 96, p.360.
2. Thomason, P.F. *Acta Metall.,* 1981, 29, p.763.
3. Thomason, P.F. *Acta Metall.,* 1985, 33, p.1087.
4. Thomason, P.F. *Acta Metall.,* 1985, 33, p.1079.
5. Hill, R. *The Mathematical Theory of Plasticity,* Clarendon Press, Oxford 1950.
6. Koiter, W.T. *Progress in Solid Mechanics* Vol. I, North Holland, 1960.
7. Kudo, H. *Int. J. Mech. Sec.,* 1960, 2, p.102.
8. Berg, C.A. *Inelastic Behaviour of Solids,* edited by M.F. Kanninen, p.171, Melgraw-Hill, New York, 1970.
9. Gurson, A.L. *J. Engng. Mater. Tech.,* 1977, 99, p.2.

10. Rudnicki, J.W. and Rice, J.R. *J. Mech. Phys. Solids,* 1975, 23, p.371.
11. Yamamoto, H. *Int. J. Fract.,* 1978, 14, p.347.
12. Wilsdorf, H.G.F. *Mat. Sci. and Eng.* 1983, 59, p.1.
13. Neumann, P. *Mat. Sci. and Eng.* 1976, 25, p.217.
14. Le Roy, G., Embury, J.D., Edwards, G., and Ashby, M.F. *Acta Metall.,* 1981, 29, p.1509.
15. Cockcroft, M.G. and Latham, D.J. *J. Inst. Metals,* 1968, 96, p.33.
16. Thomason, P.F. *Int. J. Mech. Sci.* 1969, 11, p.187.
17. Oyane, M. *Bull. J.S.M.E.* 1972, 15, p.1507.
18. Thomason, P.F. Proc. of First Int. Conf. on Technology of Plasticity, Tokyo, September 1984, published as *Advanced Technology of Plasticity 1984,* Vol.1. p.719.
19. Oyane, M. et al, *Trans Japan Soc. Mech. Engrs.,* 1973, 39, p.318.

CHAPTER 6

Ductile Fracture in Uniaxial Tension and Compression Tests

6.1 Uniform Plastic Flow in Uniaxial Tension Specimens of Cylindrical Form

The uniaxial tension specimen with a circular cylindrical gauge-length represents an extremely useful and convenient test-geometry for studying the complete process of ductile fracture in metals, giving test conditions where the full plastic stress- and strain-history can be estimated to a high degree of accuracy. With increasing tensile load F a tension specimen of initial gauge-length L_o and cross-sectional area A_o will first reach the plastic yield-point stress at which permanent deformation begins. Beyond the yield point a metal in the annealed or heat-treated state will exhibit a pronounced work-hardening effect giving a strongly increasing load, with increasing uniform plastic extension of the cylindrical gauge length L, and a corresponding uniform reduction in the cross-sectional area A, see Fig. 6.1(a); typical results for a low-carbon steel (SAE 1020) tension specimen obtained by Bluhm and Morrissey [1] are shown in Fig. 6.1(b). In this uniform mode of tensile plastic flow, constancy of volume gives the following relation between the initial ($A_o L_o$) and current (A, L) geometry of the gauge length:-

$$A_o L_o = A L . \qquad (6.1)$$

The axial principal strain ϵ_1 is therefore given by the following expression:-

$$\epsilon_1 = \ln\left(\frac{L}{L_o}\right) = \ln\left(\frac{A_o}{A}\right), \qquad (6.2)$$

and the corresponding axial principal stress σ_1 is given by:-

$$\sigma_1 = \frac{F}{A} . \qquad (6.3)$$

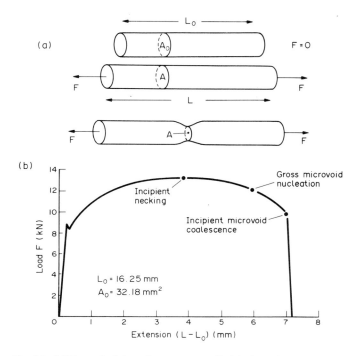

Fig. 6.1. (a) The uniaxial tension test on a cylindrical specimen showing
the initial uniform plastic state and the subsequent localised
necking effect beyond the bifurcation strain. (b) Experimental
results for a tension test on a low-carbon SAE 1020 steel specimen in a displacement-controlled testing machine. (After J.I.
Bluhm and R.J. Morrissey, Proc. 1st. Int. Conf. on Fracture, Sendai, Japan, 1965, DII-5, p.1739).

The uniform cylindrical deformation of the tension-specimen gauge length continues until the rate of work-hardening is reduced sufficiently to allow a loss of
uniqueness in the plastic velocity-field [2] and the development of a bifurcation in
the flow path. In a cylindrical tension specimen the bifurcation mode is in the form
of the well-known localised-necking effect that develops from the previous
uniform flow-field and concentrates all subsequent plastic deformation into an isolated region along the gauge length [1], Fig. 6.1(a) and (b). The point of incipient
bifurcation and necking is found to occur either at, or just beyond, the point of
maximum tensile load [1,2,3] and a simple analysis of the conditions corresponding to this critical state of deformation can be obtained following the approach of
Considére. In terms of the axial stress σ_1 and strain ϵ_1, defined by equations (6.3)
and (6.2), respectively, the axial load is $F = \sigma_1 A$ and the maximum load condition is

represented by setting $dF/d\epsilon_1 = 0$; hence the critical condition for incipient necking is:-

$$\frac{dF}{d\epsilon_1} = \frac{d(\sigma_1 A)}{d\epsilon_1} = A\frac{d\sigma_1}{d\epsilon_1} + \sigma_1\frac{dA}{d\epsilon_1} = 0 . \tag{6.4}$$

Now, from equation (6.2) it is clear that $d\epsilon_1 = -dA/A$ and thus we can rewrite the maximum load condition (6.4) in the following compact form:-

$$\frac{d\sigma_1}{d\epsilon_1} = \sigma_1 . \tag{6.5}$$

This equation states that the maximum load occurs at a strain $\epsilon_1 = \epsilon_1^u$ where the work-hardening rate has become equal to the axial stress; beyond this point the condition for uniqueness of the velocity field is no longer satisfied and a bifurcation in the plastic field can develop in the form of a localised neck [2,3]. It is interesting to note at this point that Hill's analysis of uniqueness in rigid-plastic solids [2] shows that a tension specimen subjected to a fluid hydrostatic pressure P will exhibit the same bifurcation strain ϵ_1^u, determined by equation (6.5), whatever the magnitude of P; this follows from the fact that an all-round fluid surface-pressure only changes the plastic stress-state by an additional mean-normal stress $\sigma_m = -P$. The independence of the bifurcation strain in uniaxial tension tests to wide variations in fluid hydrostatic-pressure is confirmed in the experimental work of French and Weinrich (5,6,7].

The uniaxial stress/strain relations for a wide range of metals and alloys, tested from the annealed or heat treated state, can be closely approximated by a Ramberg-Osgood power-hardening law of the following form:-

$$\sigma_1 = C\epsilon_1^m ; \tag{6.6}$$

where C is a material constant and m is the work-hardening exponent. For materials obeying the power-hardening law the maximum-load condition (6.5) gives:-

$$\frac{d\sigma_1}{d\epsilon_1} = mC\epsilon_1^{m-1} = \frac{m\sigma_1}{\epsilon_1} = \sigma_1 . \tag{6.7}$$

and thus the axial strain ϵ_1 at the incipient-bifurcation state is equal in magnitude to the strain-hardening exponent, ie. for a Ramberg-Osgood material:

$$\epsilon_1^u = m . \tag{6.8}$$

Beyond the bifurcation strain ϵ_1^u in a tension test, where localised necking develops, the so called 'true strain' and 'true stress' across the minimum neck-section can be

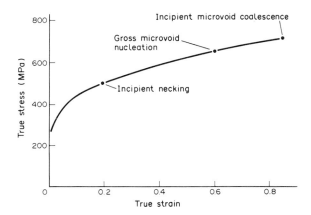

Fig. 6.2. True stress/strain curve for the SAE 1020 steel tension specimen.
(After J.I. Bluhm and R.J. Morrissey, Proc. 1st Int. Conf. on Fracture, Sendai, Japan, 1965, DII - 5, p.1739).

defined by equations (6.2) and (6.3), respectively, on the understanding that A is
now taken to be the cross-sectional area of the minimum section. These definitions
of 'true' stress and strain are of course based on the assumption that the states of
stress and strain across the minimum section are uniform. The 'true' stress/strain
curve for the SAE 1020 steel obtained by Bluhm and Morrissey [1] from their load/
extension curve (Fig. 6.1) is shown in Fig. 6.2 up to the point of incipient ductile fracture. The assumptions of uniform stress and strain over the minimum section of a
neck neglect the possibility of plastic stress and strain intensification in the central
region of the neck, and theoretical solutions to the necking problem suggest that
stress- and strain-intensification effects could be appreciable [3,7]. It is therefore
necessary to make allowances for non-uniformities in the plastic field over the
minimum neck section, throughout the post-bifurcation deformation.

6.2   Non-Uniform Plastic Flow in the Neck of a Uniaxial Tension Specimen.

Perhaps the most useful approximate analysis of the plastic stress-state in the neck
of a tension specimen is that carried out by Bridgman [8,9], which is based on the
assumption that material elements are deformed uniformly over the minimum
neck-section, thus giving a uniform state of work-hardening to a yield stress Y at a
uniform axial strain of ln (A_o/A). From this basic assumption Bridgman shows that
the stress-field equations for the minimum neck-section (Fig. 6.3) will have the following approximate forms:-

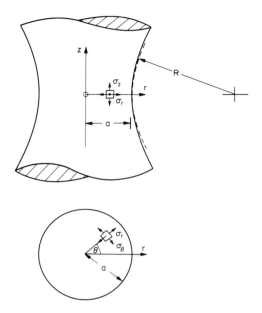

Fig. 6.3. The neck of a tension specimen showing the cylindrical coordinate system (r,θ,z) and the neck geometry parameters (a,R).

$$\frac{\sigma_z}{Y} = 1 + \ln\left(\frac{a^2 + 2aR - r^2}{2aR}\right), \qquad (6.9a)$$

$$\frac{\sigma_r}{Y} = \frac{\sigma_\theta}{Y} = \ln\left(\frac{a^2 + 2aR - r^2}{2aR}\right); \qquad (6.9b)$$

where σ_z, σ_r, σ_θ are the axial, radial and circumferential stresses, which are also
principal stresses, a is the radius of the minimum section and R is the radius of curvature of the neck surface in the longitudinal plane at the minimum section. A later
numerical study of the tensile-necking problem by Needleman [3], based on finite-
element methods, shows that the Bridgman equations form a good approximation
to the distribution of stresses across the minimum neck section, Fig. 6.4. However,
the finite-element results suggest that the mean-normal stress σ_m at the centre of
the neck $(r=0)$ from Bridgman's solution may tend to underestimate the actual
value in a real tension test; an effect that seems likely to intensify with increasing
reduction in area, Fig. 6.4.

An important feature of the Bridgman equations is that a relatively simple empirical expression is available [8,9] for the neck-geometry ratio a/R which appears to

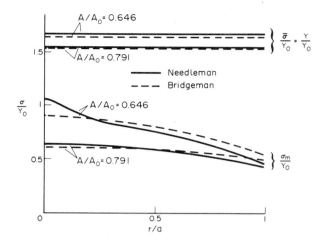

Fig. 6.4. The theoretical stress distributions at the minimum section of a tension-specimen neck from Bridgman's formula [8] and Needleman's finite-element analysis (After A. Needleman [3]).

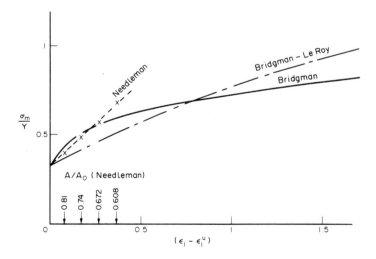

Fig. 6.5. The changing mean-normal stress σ_m at the centre of a tension-specimen neck with increasing post-bifurcation strain $(\epsilon_1 - \epsilon_1^u)$, for various numerical and empirical results [3,8,10].

be valid up to reductions in area of $\sim$ 95%, for specimens with a gauge-length to diameter ratio of about three; this empirical equation can be written in the form:-

$$\frac{a}{R} = \left[\ln\left(\frac{A_o}{A}\right) - \epsilon_1^u \right]^{\frac{1}{2}} , \qquad (6.10)$$

which indicates that the neck-geometry ratio is primarily a function of the reduction in area and the bifurcation or incipient-necking strain ϵ_1^u. Equations (6.9) and (6.10) can be used to give a closed-form expression for the variation in mean-normal stress $\sigma_m = \frac{1}{3}(\sigma_z + \sigma_r + \sigma_\theta)$ at the centre of a neck, as a tension test proceeds, and this has the form:-

$$\frac{\sigma_m}{Y} = \frac{1}{3} + \ln\left(1 + \frac{1}{2}\sqrt{\epsilon_1 - \epsilon_1^u}\right); \qquad (6.11)$$

where $\epsilon_1 = \ln (A_o/A)$ is the 'uniform' axial strain at the minimum section. This equation provides an estimate of the history of mean-normal stress σ_m at the centre of a tension-specimen neck, where σ_m has its greatest value and ductile fracture initiates [1], (cf. Chapter 1.6). Equation (6.11) can therefore be used in the subsequent theoretical estimates of ductile-fracture strains in tension tests, where it is necessary to take full account of the variation of σ_m as the necking effect develops. The variation in σ_m/Y with post-necking strain $(\epsilon_1 - \epsilon_1^u)$ from equation (6.11) is shown in Fig. 6.5 along with some numerical results from Needleman's finite element study [3]. The finite-element results in Fig. 6.5 have been modified to give σ_m/Y rather than the σ_m/Y_o normalisation in terms of the initial yield stress Y_o, obtained by Needleman, and it is clear that the two sets of results are in good agreement up to a post-necking strain equivalent to $A/A_o = 0.672$. Beyond this reduction-in-area, however the results show a wide divergence at the final numerical result for $A/A_o = 0.608$, and it is more than likely that the finite-element results are beginning to exhibit limitations of validity imposed by the rather coarse finite-element grid used in Needleman's analysis. The Bridgman equation (6.11) is also compared in Fig. 6.5 with a σ_m/Y equation introduced by Le Roy et al [10] which is based on the following empirical expression for the neck-geometry ratio:-

$$\frac{a}{R} = \kappa\left(\epsilon_1 - \epsilon_1^u\right); \qquad (6.12)$$

where κ was found to have a value $\kappa = 1.11$ for their spheroidised-steel specimens. On substituting equation (6.12) into equations (6.9) and evaluating σ_m/Y at the neck centre $(r = 0)$, we obtain the Bridgman - Le Roy results shown in Fig. 6.5. The σ_m/Y results obtained from the Le Roy et al equation (6.12) underestimate those from the Bridgman equation (6.10) at low strains, but the reverse effect occurs at high strains, Fig. 6.5. In the following work both the Bridgman and the Le Roy equation will be used to estimate the mean-normal stresses in the necks of uniaxial tension specimens.

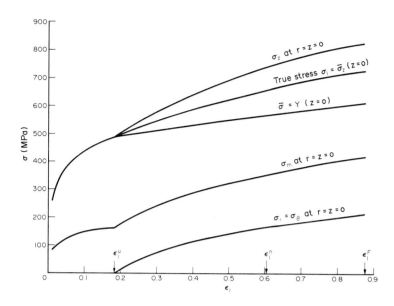

Fig. 6.6. Theoretical estimates of the changing stresses at the centre of the neck of an SAE 1020 steel tension specimen [1], using Bridgman's method [8].

6.3 The Nucleation of Microvoids in the Neck of a Uniaxial Tension Specimen

Experiments carried out by Bluhm and Morrissey [1] on a displacement-controlled tension-test machine, with a continuous ultrasonic-searching technique to monitor microvoid nucleation, show that in a wide range of materials the axial strain $\epsilon_1 = \epsilon_1^n$ at which gross microvoid nucleation occurs is well in excess of the strain at incipient necking ϵ_1^u and this is a consequence of the general need for enhanced mean-normal stress levels σ_m to initiate microvoid-nucleation mechanisms (cf. Chapter 2.1). Typical results from Bluhm and Morrissey [1] are given in Figs. 6.1(b) and 6.2 where the incipient necking strain $\epsilon_1^u = 0.182$ and gross microvoid-nucleation strain $\epsilon_1^n \approx 0.604$ are shown for the SAE 1020 steel specimens. Using the Bridgman equations (6.9) to (6.11) and Bridgman's expression for the mean axial stress $\bar{\sigma}_z$ (or 'true' stress σ_1) on the minimum section defined by [8,9]:-

$$\frac{\bar{\sigma}_z}{Y} = \left(1 + \frac{2R}{a}\right) \ln\left(1 + \frac{a}{2R}\right), \tag{6.13}$$

the results for the SAE 1020 steel in Fig. 6.2 can be replotted as shown in Fig. 6.6 where the basic uniaxial yield stress $Y = \bar{\sigma}$, true stress $\sigma_1 = \bar{\sigma}_z$, mean-normal stress

σ_m and σ_z, σ_r, σ_θ stresses at the centre of the neck ($r = 0$) are plotted against the axial tensile strain $\epsilon_1 = \ln(A_o/A)$ at the minimum neck section. These results clearly show the progressive elevating effects of both work-hardening and plastic necking constraint on the stresses in the central region of a tension-specimen neck. An interesting feature of the results in Fig. 6.6 is the fact that the average void-nucleation strain of ϵ_1^n = 0.604 for Bluhm and Morrissey's SAE 1020 steel specimen [1] corresponds to a mean-normal stress of $\sigma_m = 352$ MPa in the central neck region ($r = 0$). This result is apparently in close agreement with the Goods and Brown equation (2.5) and the associated results in Fig. 2.2 for the critical nucleation strains at various mean-normal stresses for the Fe $-$ Fe$_3$C matrix/particle system. Unfortunately, however, the results of Bluhm and Morrissey [1] do not contain sufficient information on the precise microstructural state of their tension specimens to allow definitive comparisons with the theoretical microvoid-nucleation model of Goods and Brown [11]. A set of results for plastic flow and ductile fracture in uniaxial tension specimens of various spheroidised steels, obtained by Le Roy et al [10], does contain precise information on the microstructural state of the materials and these results will now be used, in preference to the Bluhm and Morrissey results, in subsequent comparisons of experimental and theoretical values for microvoid nucleation, growth and coalescence in the neck of a tension specimen.

The tension test results of Le Roy et al [10] were obtained for the carbon steels shown in Table 6.1 which gave a wide range of volume-fractions of cementite (Fe$_3$C) particles varying from ~ 0.0167 to ~ 0.1359, and with mean particle-diameters less than 1μm which satisfies the condition for validity of the Goods and Brown void-nucleation model [11]. These tension specimens give incipient-necking strains ϵ_1^u in the range 0.24 to 0.16 and gross microvoid-nucleation strains ϵ_1^n in the range 0.62 to 0.47, Table 6.1. The nucleation-strain results for the 1045 steel specimens are plotted as the Le Roy et al results in Fig. 2.2 and are in good agreement with the Goods and Brown model for microvoid nucleation [10,11], which can be deduced from Fig. 2.2 to have the following form appropriate to a spheroidised carbon steel of $\sim 0.5\%$ carbon content:-

$$\epsilon_1^n \approx \left(\frac{1200 - \sigma_m}{1071} \right)^2 , \qquad (6.14)$$

where σ_m is the mean normal stress in MPa units and the critical stress for decohesion of the carbide/matrix interface has a value $\sigma_c \approx 1200$ MPa. Although the material constant of 1071 MPa in equation (6.14) is related to both the particle size and volume fraction (cf. equation (2.5)) the variations in these parameters for the steels in Table 6.1 are sufficiently small to allow equation (6.14) to be applied to the full set of results. Using the data of Le Roy et al [10] and the Bridgman equations (6.10), (6.11) and (6.13), the mean-normal stress σ_m at the centre of the neck at the measured nucleation strains in each specimen are given in Table 6.1; the corresponding theoretical nucleation-strain is then obtained for each specimen from equation

Table 6.1.　Summary of the results of Le Roy et al [10] for the plastic flow and ductile fracture of spheroidised steels of various carbon content.

STEEL SAE	C (wt%)	Microstructural Measurements			Strain at Incipient necking ϵ_1^u	Strain at Gross Microvoid nucleation ϵ_1^n	Strain at Ductile Fracture ϵ_1^F	Initial Yield Stress σ_{ys} (MPa)	Uniaxial Stress-Strain Relation $\sigma_1 = C\epsilon_1^m$		Theoretical Results from Bridgman's Formula when $\epsilon_1 = \epsilon_1^n$		Theoretical nucleation strain from Equation (6.14) ϵ_1^n
		Particle Volume Fraction V_f	Mean Particle Diameter $d_p(\mu m)$	Mean centre-to-centre Particle Spacing $\lambda_s(\mu m)$					C (MPa)	m	$Y=\bar{\sigma}$ (MPa)	σ_m (MPa)	
1015	0.12	0.0167	0.84	7.02	0.24	0.62	1.40	239	723	0.31	535	322	0.67
1035	0.34	0.0453	0.89	4.20	0.18	0.58 (interpolated)	1.03	267	871	0.29	612	372	0.60
1045	0.46	0.0768	0.70	2.60	0.18	0.55	0.91	302	940	0.22	716	429	0.52
1090	0.96	0.1359	0.72	1.98	0.16	0.47	0.63	464	1115	0.19	789	457	0.48

(6.14). A comparison of these theoretical and measured nucleation strains ϵ_1^n in Table 6.1 shows that they agree within $\pm$ 6% thus confirming the general validity of the Goods and Brown microvoid-nucleation model.

Once the gross microvoid-nucleation strain has been reached there are two possible modes of subsequent plastic response. The first possibility is the relatively rare case of spontaneous coalescence of the newly formed microvoids by the mechanism described in Chapters 4.5 and 5.2, where the intervoid matrix is in a plastic limit-load state as soon as the microvoids are nucleated. These conditions are only likely to occur in strongly-bonded matrix/particle systems with relatively high particle volume-fractions. The second and more common possibility is the continuing plastic growth of the microvoids to the point of incipient limit-load failure of the intervoid matrix (cf. Chapter 3.3, 3.4 and 5.3).

6.4   The Growth and Coalescence of Microvoids in the Neck of a Tension Specimen.

The individual microvoids which have been nucleated in the central region of a tension-specimen neck can be represented by the unit-cell model, shown in Figures 5.1(b) and 5.6(a) of Chapter 5, whenever the randomly-distributed microstructural particles at which the voids were nucleated are of roughly spherical form with approximately uniform diameters d_p and centre-to-centre spacings λ_s; conditions which were satisfied in the tension specimens of Le Roy et al [10]. From a purely statistical viewpoint, any point-to-point *variations* in the distributions of a particular set of particle diameters and spacings can be regarded as negligible over the *incipient* initial fracture-surface at the centre of a tension-specimen neck which will be typically $\sim$ 10 mm^2 in 'area'; the incipient fracture surface will therefore 'contain' approximately 7×10^5 particles and associated microvoids when the mean particle-diameters and centre-to-centre spacings are of the order of 1 μm and 4 μm, respectively, (see Table 6.1). Clearly with over half-a-million microvoids likely to be associated with the microvoid-coalescence process at an initial macroscopic fracture-surface, there is a sufficiently large sample of particles and voids to assume a uniform distribution of particle diameters d_p and spacings λ_s, when the particles are predominantly of spherical or equiaxed form. Hence, under these conditions, the *initial* geometry of the intervoid matrix, immediately following gross microvoid-nucleation, can be represented in terms of a single parameter $V_f \approx \pi/6$ $(d_p/\lambda_s)^3$ which is the initial volume fraction of microvoids at the nucleation strain ϵ_1^n. This of course does not rule out the possibility that, in certain materials, variations in microstructural-particle types, morphologies and spacings may be of such an extent that dual- or multiple-population models of microvoid nucleation and growth may be required to give an adequate description of the ductile-fracture process; as discussed in Chapter 2.1. However the specimens used by Le Roy et al [10] in obtaining the plastic flow and fracture results shown in Table 6.1 were of sufficient uniformity to enable a single-population model to be used in the following theoretical estimates of ductile-fracture strains, even though Le Roy et al found it necessary to use a dual-population model to explain certain features of their experimental results [10].

With the unit-cell model of a single microvoid in a large microvoid population (cf. Chapter 5 and Fig. 5.1(b) and 5.6(a)) the changing geometry of the microvoid at the centre of a tension-specimen neck, with continuing void-growth strain, can be estimated from the Rice/Tracey equations (2.16). On making the substitution $a/a_o = R_1/R_o$, $b/b_o = R_2/R_o$, $c/c_o = R_3/R_o$ and replacing ϵ_1 by the subcritical void-growth strain $(\epsilon_1 - \epsilon_1^n)$, equations (2.16) can be rewritten for uniaxial-tensile deformation ($\nu = +1$) in the form:-

$$\frac{a}{a_o} = A + B, \qquad (6.15a)$$

$$\frac{b}{b_o} = \frac{c}{c_o} = A - \frac{B}{2}; \qquad (6.15c)$$

where $A = \exp\left(D\left(\epsilon_1 - \epsilon_1^n\right)\right)$, $B = (A-1)(1+E)/D$, and $(1+E)$ and D have the values given in connection with equation (2.13); these equations are of course only valid for $\epsilon_1 \geqslant \epsilon_1^n$. The equations (2.16) and (6.15) were obtained for the case where the normalised mean-normal stress σ_m/Y is constant throughout the strain path, but the results in Fig. 6.6 show that, over the void-growth strain path from $\epsilon_1^n = 0.604$ to $\epsilon_1^F = 0.85$, σ_m/Y shows a slight increase from 0.62 to 0.67. Hence in evaluating equation (6.15) it will be assumed that σ_m/Y has a constant value amplified by a factor $G \approx 1.1$ with respect to the value of σ_m/Y at the microvoid-nucleation strain ϵ_1^n. Hence the 'constant' value of σ_m/Y for estimating the D parameter in equations (6.15) is obtained from Bridgman's equation (6.11) in the following form:-

$$\left.\frac{\sigma_m}{Y}\right|_{const.} = \left.G\frac{\sigma_m}{Y}\right|_{\epsilon_1^n} \approx 1.1\left[\tfrac{1}{3} + \ln\left(1 + \tfrac{1}{2}\sqrt{\epsilon_1^n - \epsilon_1^u}\right)\right]. \qquad (6.16)$$

It should be noted at this point that Le Roy et al [10] were able to integrate the Rice/Tracey equations (2.13) with a variable value of σ_m/Y because they evaluated the extensional growth rate $\dot{R}_1$ with respect to the major elliptical radius R_1 and the transverse growth rate $\dot{R}_3$ with respect to the minor elliptical radius R_3. Hence, effectively, Le Roy et al obtained overestimates of longitudinal growth-rates appropriate to a spherical void of radius R_1, coupled with underestimates of transverse growth-rates appropriate to a spherical void of radius R_3, Fig. 6.7(a). The present void-growth equations (2.16) and (6.15) were obtained by relating both the longitudinal and transverse growth-rates to a *mean* spherical radius given by equation (2.15); this is the preferred method of integrating the Rice/Tracey equations [12].

The results in Fig. 6.6 show that by the time the conditions for microvoid nucleation have been achieved in a tension test on a low-carbon steel specimen the work-

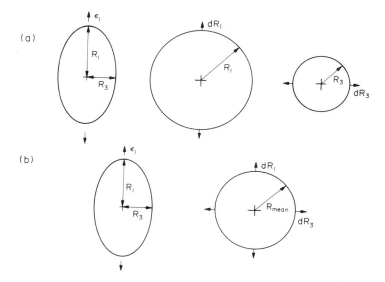

Fig. 6.7. The basis for the integration of the Rice and Tracey void-growth
equation [12] used by (a) Le-Roy et al [10] and (b) the more
appropriate mean-spherical-radius approach.

hardening rate has been reduced to a relatively small and diminishing value, and
it is therefore appropriate to use the non-hardening values of $(1 + E) = 1.667$ and
$D = 0.558 \sinh (3\sigma_m/2Y) + 0.008 \cosh (3\sigma_m/2Y)$ in the void-growth equation (6.15)
to evaluate a/a_o and b/b_o up to the point of incipient ductile fracture.

It was shown in Chapter 5.3 that the critical condition for termination of the ductile
void-growth strain, in a three-dimensional array of ellipsoidal microvoids, by plastic limit-load failure of the intervoid matrix, is given by equation (5.17). The left-
hand side (LHS) of equation (5.17) is primarily a function of the changing geometry
of the intervoid matrix and the associated plastic constraint factor [13,14], and can
be written in the form:-

$$\text{LHS} = \left(\frac{0.1}{(a/d)^2} + \frac{1.2}{(\frac{b}{b+d})^{1/2}} \right) (1 - V_f)^{-1} \left(1 - (\frac{3\sqrt{\pi} V_f}{4})^{2/3} (\frac{b}{b_o})^2 e^{(\epsilon_1 - \epsilon_1^n)} \right) ; \qquad (6.17)$$

where $(\epsilon_1 - \epsilon_1^n)$ is the subcritical void-growth strain and the various parameters are
defined in Chapter 5.3 and Fig. 5.1 and 5.6. The right-hand side RHS of equation
(5.17) is a function of σ_m/Y, at the centre of the tension-specimen neck, and from
Bridgman's equation (6.11) this can be written:-

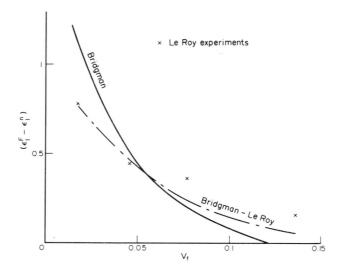

Fig. 6.8. A comparison of the theoretical and experimental void-growth strains to incipient ductile-fracture in tension tests for various spheroidised carbon steels, obtained with the equations of both Bridgman and Le Roy et al for the neck stresses. (Experimental results after G. Le Roy et al [10]).

$$\text{RHS} = 1 + \ln(1 + \tfrac{1}{2}\sqrt{\epsilon_1 - \epsilon_1^u}). \qquad (6.18)$$

Under normal conditions LHS > RHS at the void-nucleation strain ϵ_1^n and continuing plastic deformation is required to bring about sufficient void growth to reduce the plastic constraint for limit-load failure of the intervoid matrix to the point where LHS = RHS, this giving the condition for incipient ductile fracture (cf. Chapter 5.3 and Fig. 5.8). Equations (6.15) to (6.18) are too complex to allow closed-form solutions for the critical void-growth strain at incipient fracture $(\epsilon_1^F - \epsilon_1^n)$ and it is therefore necessary to evaluate the expressions for LHS and RHS numerically for increasing values of $(\epsilon_1 - \epsilon_1^n)$, until the condition LHS = RHS is reached at the incipient ductile-fracture strain $(\epsilon_1^F - \epsilon_1^n)$. These effects are similar to those illustrated in Fig. 5.8, with the exception that RHS now increases with plastic strain ϵ_1 due to the progressive development of the tension-specimen neck. Theoretical estimates of the critical void-growth strains ϵ_1^G to ductile fracture, for the Le Roy et al results [10], were made by the above method and are plotted in Fig. 6.8 as a function of the volume-fraction of cementite particles. Two sets of theoretical results are given in Fig. 6.8: in the first set of results equations (6.15) to (6.18) are evaluated using Bridgman's equation (6.10) for the a/R function at a tension-specimen neck; in the

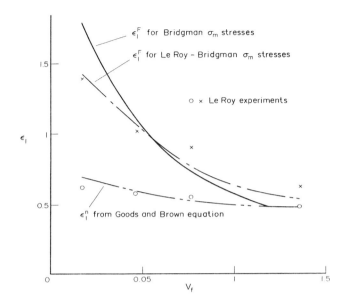

Fig. 6.9. A comparison of the theoretical and experimental void-nucleation ϵ_1^n and fracture ϵ_1^F strains in tension tests for various
spheroidised carbon steels. (Experimental results after G. Le Roy
et al [10]).

second set of results the equations are evaluated with the empirical equation (6.12)
for a/R (with $\kappa = 1.11$) which was derived by LeRoy et al from their experimental
results [10]. The results in Fig. 6.8 show a close agreement between the theoretical
void-growth strains and the experimental results when the Le Roy et al equation
(6.12) is used to give the neck-geometry parameter a/R; there is however a ten-
dency to underestimate the actual values of ($\epsilon_1^F - \epsilon_1^n$), at the higher values of V_f,
which may be the result of an experimental observation that a smaller percentage
of cementite particles initiate microvoids at higher volume-fractions of particles
[10]. The results obtained with Bridgman's equation (6.10) for the a/R parameter
tend to overestimate the void-growth strains at low V_f and underestimate these
strains at high V_f; this implies that the Bridgman empirical expression for a/R leads
to an overestimate of σ_m/Y at small necking strains, with a corresponding underestimate at higher necking strains, Fig. 6.5. An interesting comparison of the total theoretical fracture strain, which is the sum of the predicted nucleation strain from the Goods
and Brown equation (6.14) and the predicted void-growth strain from equations (6.15)
to (6.18), is made with the experimental results of Le Roy et al in Fig. 6.9; these results
show an excellent agreement between the theoretical and experimental values.

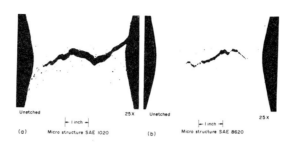

Fig. 6.10. Macrographs of the cross-sectioned neck regions of partially
fractured SAE 1020(a) and SAE 8620 (b) low carbon steel specimens. (After J.I. Bluhm and R.J. Morrissey, Proc. 1st Int. Conf.
on Fracture, Sendai, Japan, 1965, DII-5, p.1739).

6.5   The Formation and Propagation of a Ductile-Fracture Surface in the Neck of a Tension Specimen.

The theoretical ductile-fracture strains obtained in the previous section represent
the attainment of a state of *incipient* microvoid coalescence in the central region of
a tension-specimen neck, not the final state of 'cup and cone' fracture; we must
therefore begin by considering the formation of the initial macroscopic fracture-
surface at the central region of the neck. Now it was pointed out in Chapter 3.2 that
there will always be a strong tendency for the initial ductile-fracture surfaces to
form across characteristic surfaces of the plastic velocity-field, whenever they
exist. For the case of a uniaxial tensile strain-rate field ($\dot{\epsilon}_1, \dot{\epsilon}_2 = \dot{\epsilon}_3 = -\dot{\epsilon}_1/2$), similar
to that at the centre of a tension-specimen neck, characteristic surfaces which fully
satisfy condition (3.3) do not strictly exist. Nevertheless, it was shown in Chapter
3.2 that, for a uniaxial strain-rate field, surfaces do exist which contain 'characteristic directions' at an orientation angle $\psi = \pm 54.74°$ with respect to the direction of
the $\dot{\epsilon}_1$ strain-rate (cf. Fig. 3.1); thus partially satisfying condition (3.3). It is therefore
likely that the initial macroscopic fracture-surface in the neck of a tension specimen
will tend to lie parallel to these characteristic *directions*. The cross-sectional macrographs of the low-carbon steel tension specimens of Bluhm and Morrissey [1] in
Fig. 6.10 do in fact suggest that the initial fracture surface in the central neck region
is closely aligned with the local *characteristic directions* of the velocity field.

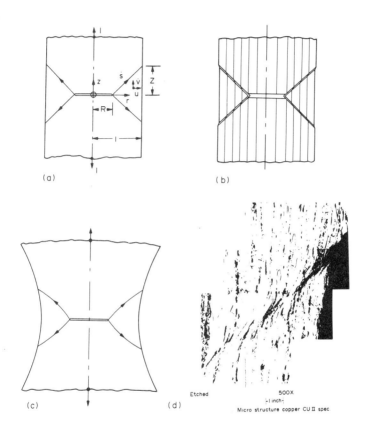

Fig. 6.11. A rigid-plastic model of a cylindrical body with an internal penny-shaped crack, showing (a) the discontinuous velocity-field and (b) the shear-band revealed by deformed vertical grid lines following a small axial plastic extension. The analogous velocity field (c) at the partially-cracked neck of a tension specimen and (d) the shear-band effect observed in the neck of a commercially-pure copper specimen (Micrograph after J.I. Bluhm and R.J. Morrissey, Proc. 1st Int. Conf. on Fracture, Sendai, Japan, 1965, DII - 5, p.1739).

After the formation of the central ductile-fracture surface in the region of high mean-normal stress σ_m, in a tension-specimen neck, the process of complete 'cup and cone' fracture requires the propagation of the initial central crack through material that has been subjected to lower mean-normal stresses (cf. equation (6.9)) and may therefore be at a substantially earlier sub-critical stage in the microvoid nucleation and growth process. Hence, the completion of the ductile-fracture process, by the *propagation* of the initial central crack through the remaining annular ligament, may require a significant additional increment of plastic deformation at the minimum neck-section for materials of very low particle volume-fraction; as observed in the tension tests of Bluhm and Morrissey on a commercially-pure copper [1], see Fig. 6.11. On the other hand, for tension tests on materials of relatively high particle volume-fraction, propagation of the initial central crack through the annular ligament may require virtually no additional increment of *bulk* plasticity; as observed in the En 3 steel specimens, Figs. 1.12 to 1.15, cf. Chapter 1.6.

An elementary model for the mechanics of additional plastic flow and ductile crack-propagation in the centrally-cracked neck of a tension specimen is shown in Fig. 6.11(a) where a central penny-shaped crack of radius R is contained in a circular bar of unit radius. If the bar is made of rigid-plastic material and subjected to unit extensional velocities at each end, a triangular-shaped plastic zone of axial length 2Z can develop with the following kinematically-admissible velocity field [15]:-

$$\dot{u} = -\left(\frac{1-R}{2Z}\right)\left(\frac{R+r}{r}\right) , \tag{6.19a}$$

$$\dot{v} = \frac{z}{r}\left(\frac{1-R}{2Z}\right) \tag{6.19b}$$

$$\dot{s} = \left(\frac{R+r}{2r}\right)\sqrt{\left(\frac{1-R}{Z}\right)^2 + \left(\frac{r-R}{r+R}\right)^2} , \tag{6.19c}$$

where $\dot{u}$ and $\dot{v}$ are the radial and axial velocities, respectively, at a point (r,z) in the plastic annulus, and $\dot{s}$ is the tangential velocity-discontinuity on the top and bottom boundaries of the plastic zone. Although this kinematically-admissible velocity field does not necessarily have an associated statically-admissible stress field, the fact that it satisfied the velocity boundary conditions and plastic-incompressibility requirements suggests that it will represent a good approximation to the potential plastic-flow field in the annular ligament of a tension-specimen neck, prior to complete 'cup and cone' fracture. For a small increment of plastic deformation in the centrally-cracked model (Fig. 6.11(a)) the velocity-field equations (6.19) give the distorted grid pattern shown in Fig. 6.11(b), where the intense shear bands at the top and bottom surfaces of the plastic zone coincide with the locations of the velocity-discontinuity $\dot{s}$ (equation (6.19c)). In a real tension specimen the neck profile is not cylindrical and so the plastic annular ligament will have slightly curved upper and lower boundaries as shown schematically in Fig. 6.11(c). In materials which require large void-growth strains to initiate ductile void-coalescence, substantial shear-band formation is often observed on the plastic ligament boundaries before the final propagation of the central crack along these bands of intense deformation and microstructural damage. An example of shear-band formation in the annular ligament of Bluhm and Morrissey's commercially-pure copper tension specimen is shown in Fig. 6.11(d) [1]. With materials of higher particle volume-fraction, similar to the En3 steel specimens described in Chapter 1.6, the central ductile crack can cleary propagate across the annular ligament with no observable shear-band formation (see Fig. 1.15).

The frequently observed shear-band formations in the ligaments of partially-cracked tension specimens of high ductility have been responsible in the past for a fundamental misunderstanding of the basic mechanism of ductile fracture, leading to attempts to represent the ductile-fracture process by intense shear-band models (cf. Chapter 3.5 and 3.6). It needs to be emphasised that in most processes involving ductile fracture the *initial* ductile-fracture surface is formed in a region of high plastic constraint where intense shear-band formation is excluded by compatibility requirements (cf. Chapter 3.3 to 3.6). Once the *initial* ductile-fracture surface has developed, however, the degree of plastic constraint is generally reduced to a level where intense shear-deformation bands can subsequently develop and thereby promote the *propagation* of the ductile-fracture surface. However, the basic ductile-fracture mechanisms of microvoid nucleation, growth and coalescence must then be repeated *within* the intense shear-bands.

6.6   The Effect of a Fluid Hydrostatic Pressure on the Ductile Fracture Process in Tension Specimens.

The general effects of a fluid hydrostatic pressure P on the microvoid nucleation, growth and coalescence mechanisms were considered in Chapter 4.1 and 4.2, and in this section we consider briefly the effects of P on the tension test. When a tension test is carried out under a large fluid-hydrostatic pressure plastic flow will occur in exactly the same mode as that under zero fluid pressure since the stress state in any element of a tension specimen will be changed only by a mean-normal stress equal to $-P$. In particular, as shown theoretically by Hill [2] and confirmed experimentally by French and Weinrich [4-6], the bifurcation strain ϵ_1^u at incipient necking will be the same for all magnitudes of P and the subsequent mean-normal stresses at the neck centre will be given by the Bridgman equation (6.11) for σ_m reduced by a value equal to P; ie ($\sigma_m + P$), where compressive values of P are taken to be algebraically negative. It follows from the modified Goods and Brown equation (4.2) for the microvoid nucleation strain ϵ_1^n that a large superimposed fluid-pressure P will considerably increase the nucleation strain. In addition the results in Fig. 4.2 show that large compressive values of P will severely restrict the

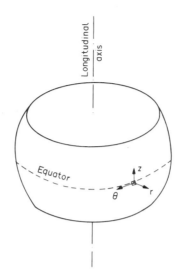

Fig. 6.12. Reference axes for an element at the equatorial free-surface of
a barrelled compression specimen.

extent of microvoid growth for a given void-growth strain $(\epsilon_1 - \epsilon_1^n)$. Finally, the conditions for microvoid coalescence (equations (4.3) and (4.4)) show that large compressive values of P tend to inhibit the process of microvoid coalescence and thus greatly extend the void-growth strain to incipient ductile fracture. Hence a large fluid hydrostatic pressure tends to suppress every stage in the ductile-fracture process of nucleation, growth and coalescence of microvoids, thus giving greatly enhanced ductility [4 - 6].

6.7   Plastic Flow in Uniaxial Compression Specimens of Cylindrical Form.

A useful alternative experimental method for studying the ductile fracture problem is based on the uniaxial compression of a cylindrical specimen between flat compression-platens [16]. Compression specimens of initial height H_o and diameter D_o can be compressed between platens that carry various types of interface lubrication giving friction conditions between the specimens and platens varying from approximately zero-friction, where the specimen remains cylindrical and exhibits only homogeneous deformation, to unlubricated conditions where approximately 'sticking-friction' effects cause a severe non-uniform 'barrelling' mode of deformation, Fig. 6.12. Experiments based on the measurements of deformed grids on the surface of pure aluminium compression specimens [16] show that 'barrelling'

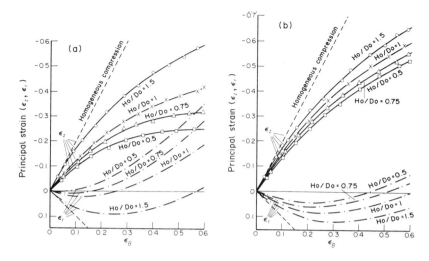

Fig. 6.13. The relationships between the principal strains at the equatorial surface of compression specimens with (a) unlubricated platens, (b) molybdenum disulphide lubrication. (After P.F. Thomason [16]).

effects can lead to the development of highly tensile mean-normal stresses σ_m on the barrelled surface of compression specimens, which achieve their highest magnitudes on the equatorial free surface, Fig. 6.12. The equatorial free surface is thus the region where ductile fracture initiates and the barrelled compression specimen therefore has the advantage over the uniaxial tension specimen of allowing direct non-destructive observations of the progressive ductile-fracture process. Typical experimental results for the axial ϵ_z, radial ϵ_r and circumferential ϵ_θ plastic principal strains at the equatorial surface of compression specimens of various H_o/D_o ratio are shown in Fig. 6.13(a) for unlubricated platens and Fig. 6.13(b) for platens lubricated with a solid film of molybdenum disulphide. These results show the increasingly strong departure from homogeneous compression with increase in the compression-platen friction condition, an effect which is intensified as the initial height-to-diameter ratio H_o/D_o of the compression specimen is reduced [16].

The results in Fig. 6.13 give the plastic-strain histories at the equatorial free surface of the compression specimens and the gradients of the graphs, at any particular circumferential strain ϵ_θ, give the ratios of the plastic strain-increments $d\epsilon_z/d\epsilon_\theta$ and $d\epsilon_r/d\epsilon_\theta$. Now the incompressibility equation (1.1) for plastic strain-increments shows that the $d\epsilon_z/d\epsilon_\theta$ and $d\epsilon_r/d\epsilon_\theta$ ratios are not independent and can be rep-

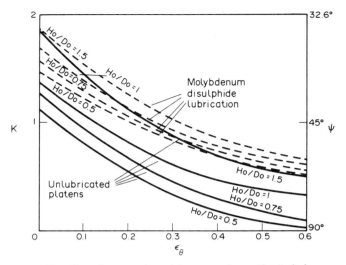

Fig. 6.14. The effect of progressive axial compression on the strain-increment ratio K at the equatorial surface for both unlubricated platens and molybdenum disulphide lubrication. Also showing the
corresponding orientation angle ψ of the velocity-field characteristics. (After P.F. Thomason [16]).

sented by the ratio:-

$$K = -\frac{d\epsilon_z}{d\epsilon_\theta} = \left(\frac{d\epsilon_r}{d\epsilon_\theta} + 1\right),\tag{6.20}$$

and thus K uniquely defines the ratios of the plastic strain-increments; the values
of K from the results in Fig. 6.13 are shown in Fig. 6.14. Since the elastic strains are
negligible in comparison to the plastic strains in the compression tests, we can
adopt the Levy-Mises flow rule [9] in place of the Prandtl-Reuss equations (1.22)
and this can be written in the following form [9,16]:-

$$d\epsilon_r = \frac{d\lambda}{3}\left[2\sigma_r - \sigma_z - \sigma_\theta\right],\tag{6.21a}$$

$$d\epsilon_\theta = \frac{d\lambda}{3}\left[2\sigma_\theta - \sigma_z - \sigma_r\right],\tag{6.21b}$$

$$d\epsilon_z = \frac{d\lambda}{3}\left[2\sigma_z - \sigma_r - \sigma_\theta\right],\tag{6.21c}$$

where $d\lambda$ is the scalar factor of proportionality related to the state of work-hardening.

The equations (6.21), together with the appropriate form of the von Mises yield criterion from equation (1.16), can now be solved to give the equatorial surface stresses $(\sigma_z, \sigma_\theta, \sigma_m)$ as functions of the plastic strain-increment ratio K [16]:-

$$\frac{\sigma_z}{Y} = \frac{1 - 2K}{\sqrt{3(K^2 - K + 1)}} , \tag{6.22a}$$

$$\frac{\sigma_\theta}{Y} = \frac{2 - K}{\sqrt{3(K^2 - K + 1)}} , \tag{6.22b}$$

$$\frac{\sigma_m}{Y} = \frac{1 - K}{\sqrt{3(K^2 - K + 1)}} , \tag{6.22c}$$

where Y is the uniaxial yield stress at the current equivalent strain $\bar{\epsilon}$ which is defined by [9] :-

$$\bar{\epsilon} = \int d\bar{\epsilon} = \sqrt{\tfrac{2}{3}} \int (d\epsilon_r^2 + d\epsilon_\theta^2 + d\epsilon_z^2)^{\frac{1}{2}} . \tag{6.23}$$

On evaluating equation (6.23) by numerical methods and using experimental values of Y from the basic uniaxial stress/strain curve [9], the results in Fig. 6.14 can be used to obtain the equatorial surface stresses from equations (6.22) and these are shown in Fig. 6.15 for a compression specimen of initial geometry $H_o/D_o = 1.5$. These results clearly show that with increasing amounts of platen friction, the increasing barrelling effect leads to the development of highly tensile mean-normal stresses σ_m at the equatorial surface, thus promoting the conditions for ductile fracture.

6.8   Ductile Fracture in Uniaxial Compression Specimens of Cylindrical Form

An experimental study of the ductile-fracture process, in 1% carbon steel compression specimens, has been carried out [17] using the methods described above to estimate the complete plastic stress and strain history at the equatorial free-surface where ductile fracture initiates. The compression tests were carried out with various platen lubrication conditions giving a wide range of platen-friction effects varying from low friction to virtually sticking friction. The variations in the principal strains $(\epsilon_z, \epsilon_r, \epsilon_\theta)$ at the equatorial surfaces of the 1% carbon-steel compression specimens, of initial height-to-diameter ratios $H_o/D_o = 1.5$ and $H_o/D_o = 0.9$, are shown in Fig. 6.16 for the two cases of unlubricated platens and molybdenum disulphide solid-film lubrication. These experimental results were used to evaluate the σ_θ and σ_m stress histories at the equatorial surfaces of the barrelled compression specimens using equations (6.22) and (6.23) [17], and the results are given in Fig. 6.17(a) for $H_o/D_o = 0.9$ and Fig. 6.17(b) for $H_o/D_o = 1.5$.

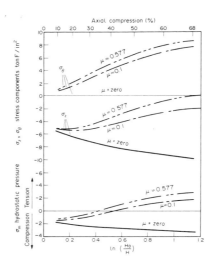

Fig. 6.15. The effect of progressive axial compression on the principal stresses at the equatorial free surface, for specimens of initial geometry $H_o/D_o = 1.5$ and various platen-friction μ conditions. (After P.F. Thomason [16]).

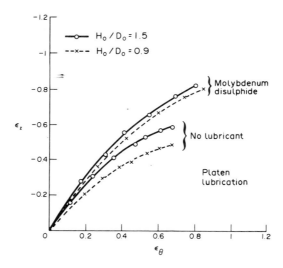

Fig. 6.16. The relationships between the principal strains at the equatorial surface of the spheroidised 1% carbon steel compression specimens. (After P.F. Thomason [17]).

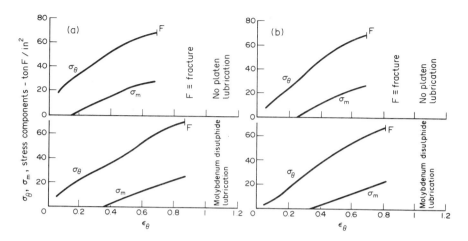

Fig. 6.17. The effect of progressive axial compression on the σ_θ and σ_m stress components at the equatorial surface of the spheroidised 1% carbon steel specimens, with various platen friction conditions, and (a) $H_o/D_o = 0.9$, (b) $H_o/D_o = 1.5$. (After P.F. Thomason [17]).

The 1% carbon steel compression specimens where tested in the spheroidised-annealed state and the microstructure consisted predominantly of small spheroidal Fe_3C carbide particles of $\sim 1\mu m$ diameter; in addition the ferrite matrix contained a much smaller number of larger carbide particles of the order of $\sim 5\mu m$ diameter [19]. The approximate volume fraction of carbides in the matrix of the compression specimens obtained by assuming that all the carbon is contained in the form of Fe_3C carbides, is of the order of 0.155 and it is therefore slightly higher than that of the SAE 1090 tension specimens used in the work of Le Roy et al [10] (Fig. 6.8 and 6.9). The results in Fig. 6.9 show that at high volume-fractions of carbides ($V_f \approx 0.15$) the ductile-fracture strain is primarily determined by the microvoid-nucleation strain ϵ_1^n because the subsequent void-growth strain ($\epsilon_1^F - \epsilon_1^n$) to incipient microvoid coalescence (Fig. 6.8) is likely to be very small or even zero. It is interesting therefore to use the compression-test results in Fig. 6.17 to obtain the theoretical microvoid-nucleation strains ϵ_θ^n, using the Goods and Brown equation (6.14) for $Fe - Fe_3C$ systems (where the maximum principal strain ϵ_1 is now replaced by ϵ_θ [11]), and compare them with the ductile-fracture strains ϵ_θ^F at the equatorial surfaces. The compression test results are given in Table 6.2 in the form of the measured circumferential strain ϵ_θ^F at fracture and the corresponding estimated value of the mean normal stress σ_m^F; this value of $\sigma_m \approx \sigma_m^F$ is then substituted in the Goods and Brown equation (6.14) to give the estimated microvoid-nucleation strains ϵ_θ^n. It is clear from the results in Table 6.1 that the differences between

Table 6.2   Uniaxial compression tests on a spheroidised 1% carbon steel [17].
(Chemical analysis :- 1%C, 0.4%Mn, 0.2% Si, 0.04%S, 0.04%P).

H_o/D_o	Platen Lubricant	Fracture Strain ϵ_θ^F	Mean-Normal Stress at Fracture σ_m^F (MPa)	Theoretical Void-Nucleation Strain ϵ_θ^n	Void-Growth Strain at Fracture $(\epsilon_\theta^F - \epsilon_\theta^n)$
1.5	Non	0.68	425	0.52	0.16
1.5	M_oS_2	0.90	340	0.64	0.16
0.9	Non	0.67	432	0.51	0.16
0.9	M_oS_2	0.86	401	0.56	0.30

the fracture strains and nucleation strains $(\epsilon_\theta^F - \epsilon_\theta^n)$, which represents the void-growth strain to fracture, are quite small and of the order $0.15 \sim 0.3$, as expected for a material having a particle volume fraction $V_f \approx 0.155$. This comparison of theoretical and experimental results is particularly good when it is considered that the ductile fracture strains ϵ_θ^F in Fig. 6.17 were measured at the appearance of a fully-developed macroscopic ductile crack on the barrelled surface; on the other hand microscopic ductile cracking was observed in the 1% carbon steel compression specimens at circumferential strains ϵ_θ about 0.1 lower than ϵ_θ^F [17]. These results indicate that at sufficiently high particle volume-fractions the ductile-fracture strains can be virtually nucleation-controlled, because once the microvoids have been nucleated only very small additional void-growth strains are required to bring about the conditions for microvoid coalescence by plastic limit-load failure of the intervoid matrix (cf. Chapter 4.5).

For compression tests on specimens of lower particle volume-fraction than those described above the nucleation strain will of course be followed by a large void-growth strain and this will increase in magnitude as the particle volume fraction is reduced, in an analogous manner to that in uniaxial tension tests, Fig. 6.6 and 6.7. Theoretical estimates of void-growth strains $(\epsilon_\theta^F - \epsilon_\theta^n)$ to ductile fracture can be obtained in a similar manner to the tensile test results in Section 6.4, there is however a great shortage of published results for ductile fracture in compression tests on specimens of low particle V_f, which contain sufficient additional details of chemical composition and microstructural state to allow adequate quantitative comparisons with theoretical results.

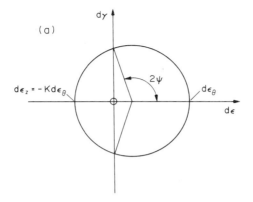

Fig. 6.18(a). The Mohr strain-increment circle showing the construction
defining the orientation ψ of the characteristics.

6.9 The Formation of Ductile-Fracture Surfaces at the Equatorial Surface of Barrelled Compression Specimens.

When ductile fracture occurs at the equatorial surface of a barrelled compression specimen the *initial* surface of microvoid coalescence is immediately observable and, as pointed out in Chapter 3.2, there will always be a strong tendency for the fracture surface to develop along a characteristic surface of the local plastic velocity-field. For the equatorial surface of the barrelled compression specimen the condition of zero extensional strain-rate in the characteristic directions is shown in Fig. 6.18(a) and this leads to the following expression for the orientation ψ of the characteristics, with respect to the direction of maximum principal strain ϵ_θ, in terms of the equatorial strain-increment ratio $K = -d\epsilon_z/d\epsilon_\theta$:-

$$\psi = \pm \tfrac{1}{2} \cos^{-1} \left(\frac{K-1}{K+1} \right). \tag{6.24}$$

The orientation angles ψ of the characteristics at the equatorial surface of aluminium compression specimens [16] are shown on the right-hand axis of the graphs in Fig. 6.14, for the corresponding values of the K ratios on the left-hand axis. These results show that, with progressive compression and barrelling, the characteristics change from an orientation of $\psi = \pm 32.6°$ at K = 2 through $\psi = \pm 45°$ at K = 1 to $\psi = \pm 90°$ at K = 0. Most compression tests reach the point of incipient ductile-fracture with characteristics orientated in the range $40° < |\psi| < 55°$ and typical ductile fractures on barrelled compression specimens are shown in Fig. 6.18(b), where $\psi = 45°$, and Fig. 6.18(c) where $\psi = 53°$. With very high platen-friction

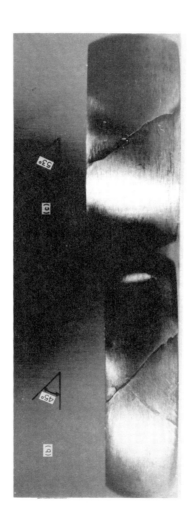

Fig. 6.18(b) and (c). Two high-carbon steel compression specimens of differing geometry and platen-friction conditions with equatorial ductile-fracture cracks at orientations of (b) $\psi \approx 45°$ and (c) $\psi \approx 53°$.

μ and large percentage compression, ductile fractures along characteristic orientations of $\psi \approx 90°$ have been observed on the barrelled surfaces of steel compression
specimens by Kudo and Aoi [18]. It is of course only the *initial* fracture surface in the
region of the equator of barrelling that is likely to be aligned with the characteristic
direction ψ; once the initial fracture surface is formed the subsequent propagation
of the ductile crack across the barrelled surface can show a substantial deviation
from the original ψ orientation, see Fig. 6.18(c).

REFERENCES

1. Bluhm, J.I. and Morrissey, R.J., *Proc 1st Int. Conf. on Fracture,* Sendai, Japan
    1965, DII - 5, p.1737.
2. Hill, R., *J. Mech. Phys. Solids,* 1957, 5, p.153.
3. Needleman, A., *J Mech. Phys. Solids,* 1972, 20, p.111.
4. French, I.E., and Weinrich, P.F., *Acta Metall,* 1973, 21, p.1533.
5. French, I.E., and Weinrich, P.F., *Scripta Metall,* 1974, 8, p.87.
6. French, I.E., and Weinrich, P.F., *Metallurgical Trans,* 1975, 6A, p.785.
7. Thomason, P.F., *Int. J. Mech. Sci., 1969, 11, p.481.*
8. *Bridgman, P.W., Studies in Large Plastic Flow and Fracture,* McGraw Hill, New
    York, 1952.
9. Hill, R., *The Mathematical Theory of Plasticity,* Clarendon Press, Oxford, 1950.
10. Le Roy, G., Embury, J.D., Edwards, G., and Ashby, M.F., *Acta Metall,* 1981, 29,
    p.1509.
11. Goods, S.H., and Brown, L.M., *Acta Metall,* 1979, 27, p.1.
12. Rice, J.R., and Tracey, D.M., *J. Mech, Phys. Solids,* 1969, 17, p.201.
13. Thomason, P.F., *Acta Metall,* 1985, 33, p.1079.
14. Thomason, P.F., *Acta Metall,* 1985, 33, p.1087.
15. Kudo, H., *Int. J. Mech. Sci.,* 1960, 2, p.102.
16. Thomason, P.F., *Int. J. Mech. Sci.,* 1968, 10, p.501.
17. Thomason, P.F., *Int. J. Mech. Sci.,* 1968, 11, p.187.
18. Kudo, H., and Aoi, K., *J. Japan Soc. Tech. Plast,* 1967, 8, p.17.

CHAPTER 7

Ductile Fracture in Metalworking
and Metalforming Processes

7.1 Ductility in Upsetting and Heading Processes

A large proportion of industrial metalworking processes are of the upsetting or heading type in which a bolt or similar component is produced by locally increasing the diameter of a bar at a chosen position along its length [1-5]; a typical two-stage heading operation, to produce a component that can be subsequently formed into a bolt head, is shown in Fig. 7.1. It is clear from a comparison of Fig. 7.1 and the barrelled compression specimen in Fig. 6.12 that the plastic flow and fracture problems in upsetting and heading operations are basically similar to those observed in the uniaxial compression of cylindrical specimens, which were considered in detail in Chapter 6.7 to 6.9. It might be presumed therefore that the methods of estimating ductility in uniaxial compression specimens, outlined in Chapter 6.8, could be applied directly to the problem of establishing the limits of ductility in upsetting and heading processes. Unfortunately, however, the limit of ductility in an industrial upsetting or heading operation is usually determined by the presence of longitudinal defects on the surface of the heading wire or bar stock, which tend to open up during the forming process and cause localised surface fractures at a much earlier stage in the process than would occur with defect-free surfaces (4). The surface defects are developed on the heading-wire during the wire-production process and cannot be eliminated by any economically viable process (1,2]; good quality wire or bar stock for upsetting or heading operations is usually processed to limit the depth of the existing surface defects to less than ~ 0.005 in [6]. It follows from this that methods of estimating ductility in upsetting and heading operations must always take into account the presence, to a greater or lesser extent, of longitudinal surface defects on the cylindrical surfaces of the wire or bar stock.

An experimental study of the effects of longitudinal surface defects, in the form of

162

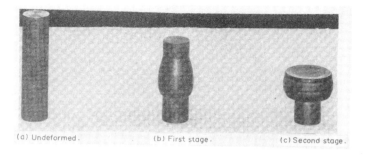

(a) Undeformed.   (b) First stage.   (c) Second stage.

Fig. 7.1. A two-stage heading process. (After P.F. Thomason, Proc. Instn.
Mech. Engrs., 1969, 184, p.875).

artificial machined-grooves, on the ductility of uniaxial compression specimens
has been carried out for a 1% carbon-steel material [2], in an attempt to simulate
the effects observed in heading and upsetting processes (1,7). The compression
specimens were machined with grooves of the following depths:- 0.001 in, 0.003 in
and 0.005 in, and a series of compression tests with various platen-friction conditions and specimen geometries were carried out to the point of ductile fracture at
the groove root. A typical set of results for compression specimens of initial height-
to-diameter ratio of 1.5 are shown in Fig. 7.2, where the circumferential strains ϵ_θ
for defect-free specimens are compared with the mean groove-strain ϵ_w which is
approximately equal to the strain at the groove root [2]. A typical groove-root fracture is shown in Fig. 7.3. These results show a severe reduction in compression-
specimen ductility with increasing groove depth and the effect is accentuated with
increasing platen/specimen friction. With unlubricated platens a defect-free specimen can be compressed by approximately 72% before fracture, but with a defect
of 0.001 in depth fracture occurs at $\sim$ 51% compression and a defect of 0.005 in
depth gives fracture at $\sim$ 45% compression. A reduction in platen/specimen friction
by interface lubrication can increase the compressive ductility to a significant
extent but the surface grooves still give a severely reduced ductility which
increases in magnitude with groove depth, Fig. 7.2.

A useful general method of assessing the effects of longitudinal surface defects on
the ductility of heading wire and bar stock can be derived from results similar to
those in Fig. 7.2 for a particular specimen geometry H_o/D_o and platen friction condition. The first point to note is that experiments show that, for particular values of
H_o/D_o and platen friction, the localised plastic flow at grooves of 0.001, 0.003 and
0.005 in depth on specimens of 0.25 in diameter closely correspond to the localised

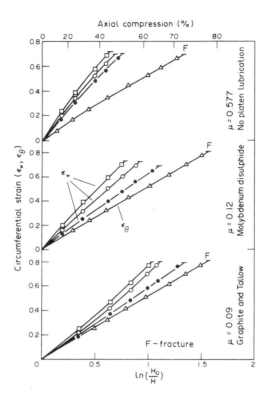

Fig. 7.2. The circumferential strains at the equatorial surface ϵ_θ and groove roots ϵ_w of spheriodised 1% carbon steel compression specimens of initial geometry $H_o/D_o = 1.5$ and various platen friction μ conditions. (After P.F. Thomason [2]).

flow at grooves of 0.002, 0.006 and 0.010 in depth on specimens of 0.5 in diameter [2]. Hence, the results in Fig. 7.2 for 0.25 in diameter specimens can be used to represent approximately the plastic flow and fracture effects in specimens of differing initial diameters D_o if the ϵ_w curves are taken to represent the initial groove depth to diameter ratio d_o/D_o. Taking now the results in Fig. 7.2 for platen lubrication by molybdenum disulphide as an example, it can be seen from the approximate linear nature of the results that the extrapolated ratio of the 'groove-root' strain $\sim\epsilon_w$ to the circumferential strain ϵ_θ on a specimen without grooves will have the following approximately constant values throughout the compression test:- $\epsilon_w/\epsilon_\theta \approx 1.22$, 1.51, 1.82 for $d_o/D_o = 0.004, 0.012, 0.020$, respectively.

Fig. 7.3. The root of a longitudinal groove in a 1% carbon steel compres-
          sion specimen, showing the development of ductile fracture (X
          35). (After P.F. Thomason [2]).

These values give a linear relation between the groove strain-concentration ratio
$\epsilon_w/\epsilon_\theta$ and the normalised groove depth d_o/D_o in the form:-

$$\frac{\epsilon_w}{\epsilon_\theta} = 1 + 42.66\frac{d_o}{D_o} , \qquad\qquad (7.1)$$

where the constant of 42.66 represents the particular initial geometry of $H_o/D_o = 1.5$
and molybdenum-disulphide platen lubrication. It does not follow that all combinations of specimen geometry and friction condition will produce a simple linear
relation between $\epsilon_w/\epsilon_\theta$ and d_o/D_o, similar to equation (7.1), but we proceed to
develop in principal a general equation for heading ductility on the basis of this
form of equation.

The results in Fig. 7.4 show an approximately linear relation between ϵ_θ and ln $(H_o/$
$H)$ which has the following form for the case of molybdenum-disulphide lubrication:-

$$\epsilon_\theta = 0.516 \ln\left(\frac{H_o}{H}\right) . \qquad\qquad (7.2)$$

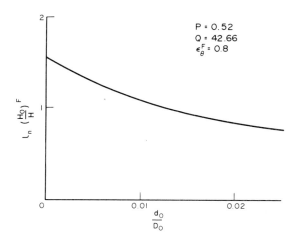

Fig. 7.4. The effect of longitudinal surface defects on the ductility of cylindrical bars in heading and upsetting processes.

Combining equations (7.1) and (7.2), with the geometry- and friction-dependent constants replaced by P ($\equiv$ 0.516) and Q ($\equiv$ 42.66), we obtain the following general expression for the axial compression corresponding to a particular groove-root strain ϵ_w:-

$$\ln\left(\frac{H_o}{H}\right) = \frac{\epsilon_w}{P\left(1 + Q\frac{d_o}{D_o}\right)} \qquad (7.3)$$

The results in Fig. 7.2 and in previous work [2] suggest that groove-root fracture (Fig. 7.3) will occur when ϵ_w is approximately equal to the fracture strain ϵ_θ^F measured on the surface of a specimen without grooves, ie., $\epsilon_w^F \approx \epsilon_\theta^F$. Hence, the axial upsetting or heading that can be carried out on a wire or bar with longitudinal surface defects is obtained from equation (7.3) by setting $\epsilon_w = \epsilon_\theta^F$ to give:-

$$\ln\left(\frac{H_o}{H}\right)^F = \frac{\epsilon_\theta^F}{P\left(1 + Q\frac{d_o}{D_o}\right)} \qquad (7.4)$$

where P and Q are constants for a particular set of geometry and friction conditions and must be obtained by experiment [2]. This equation is shown schematically in Fig. 7.4 for representative values of P, Q and ϵ_θ^F, where the strong reduction in heading ductility resulting from the presence of relatively shallow longitudinal surface defects is clearly indicated.

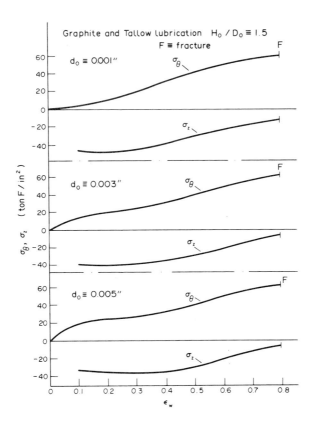

Fig. 7.5. The relationships between the stresses and circumferential strain ϵ_w at the root of longitudinal grooves on uniaxial compression specimens. (After P.F. Thomason [2]).

When applying ductility-limit equations to heading and upsetting processes, similar to equation (7.4), it needs to be understood that, for a given depth of longitudinal surface-defect on the heading wire, heat-treatment processes can in some cases reduce heading ductility and in other cases improve it [1]. These effects depend on the particular material and the amount of cold-drawing of the heading wire since the last annealing process. It should also be noted that upsetting and heading processes carried out at elevated temperatures (ie., warm-working) can in the case of steels give severely reduced ductility in the 'blue-brittle' temperature range 150°C to 350°C (cf. Chapter 4.3 and 4.4) and greatly enhanced ductility at temperatures of the order ~550°C [8].

Table 7.1. Uniaxial compression tests on a spheroidised 1% carbon steel, with longitudinal grooves on the specimen surfaces [2].

Groove Depth (ins)	Groove Root Fracture Strain ϵ_w^F	Mean-Normal Stress at Fracture σ_m^F (MPa)	Theoretical Void-Nucleation Strain ϵ_w^n	Void-Growth Strain at Fracture $(\epsilon_w^F - \epsilon_w^n)$
0.001	0.8	262	0.77	0.03
0.003	0.8	293	0.72	0.08
0.005	0.8	299	0.71	0.09

At this point it is interesting to examine the results for ductile fracture at the root of longitudinal grooves in 1% carbon steel compression specimens [2], Fig. 7.2, in terms of the fundamental mechanisms of microvoid nucleation, growth and coalescence in uniaxial compression specimens (cf. Chapter 6.7 to 6.9). Results are available [2] for the variation in the axial σ_z and circumferential σ_θ stresses at the root of grooves in the 1% carbon steel compression specimens of $H_o/D_o = 1.5$, with graphite and tallow lubrication, and are shown in Fig. 7.5. Now it was pointed out in Chapter 6.8 that the void-growth strains from incipient microvoid nucleation to incipient microvoid coalescence are likely to be very small in 1% carbon steels due to the very high volume-fraction of Fe_3C carbides and resulting microvoids; thus ductile fracture in this type of steel may be regarded as virtually nucleation-control-led. We can therefore use the results in Fig. 7.5 to estimate the mean-normal stress σ_m^F at groove-root fracture and then use the Goods and Brown equation (6.14) for $Fe - Fe_3C$ systems to estimate the microvoid-nucleation strain ϵ_w^n at the groove root. These results are tabulated in Table 7.1 where the groove-root fracture strain ϵ_w^F, mean-normal stress σ_m^F and nucleation strain ϵ_w^n are given for each groove depth. The results show that the theoretical microvoid-nucleation strains are only slightly smaller than the fracture strains, giving void-growth strains to fracture ($\epsilon_w^F - \epsilon_w^n$) of 0.03, 0.08 and 0.09 for groove depths of 0.001, 0.003 and 0.005 in, respectively. These void-growth strains to fracture compare with values of the order 0.16 $\sim$ 0.3 for a similar material tested without longitudinal grooves on the compression specimen surfaces, cf. Table 6.2. However, the two sets of results can be regarded as completely consistent because the groove machining process is likely to produce some localised groove-root damage which would promote the microvoid nucleation process.

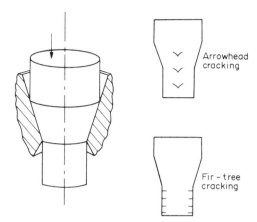

Fig. 7.6. The free-extrusion process, showing the 'arrowhead' and 'fir-tree' modes of fracture. (After P.F. Thomason, Proc. Instn. Mech. Engrs., 1969/70, 184, p.896).

7.2   Ductility in Extrusion and Drawing Processes.

The free-extrusion process through conical dies (Fig. 7.6(a)) is an important industrial cold-working process which is basically similar to both wire-drawing and hydrostatic-extrusion processes. Common forms of ductile fracture occuring at the forming limits of these processes have the appearance of internal 'chevron' or 'arrowhead' cracking [9], as shown in Fig. 7.6. In the present section we consider in detail only the problem of internal arrowhead cracking, however, the methods developed below apply in principle also to the problem of fir-tree cracking.

A typical axisymmetric slip-line field in the plastic deformation zone of the free-extrusion process is shown in Fig. 7.7 for the case of zero die-face friction. The equilibrium equations along the α and β slip lines for a material obeying the Tresca yield criterion [10] have been integrated numerically, with the aid of a modified Haar-Karman hypothesis [9,10], to give the mean-normal stress σ_m at the axis of symmetry of the plastic zone ($r = 0$, $z = 0$) for various die half-angles and logarithmic reductions in area ln (A_o/A), and typical results are shown in Fig. 7.8. These results show that very high tensile mean-normal stresses σ_m can be generated on the axis of symmetry of a free-extrusion process, and the intensity of σ_m increases with both increasing die angle and reducing extrusion ratio A_o/A. With conical dies of relatively small included angle the logarithmic reduction in area ln (A_o/A) is approximately equal to the axial tensile plastic strain $\epsilon_1 \equiv \epsilon_z$ encountered by a material element flowing through the plastic zone along the axis of symmetry [9]. Hence, the results in Fig. 7.8 give both the mean-normal stress σ_m and the total

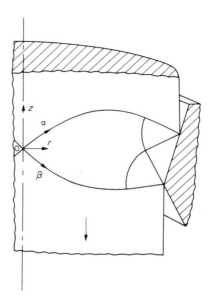

Fig. 7.7. The plastic deformation zone for free extrusion, showing the axisymmetric slip-lines in the meridian plane. (After P.F Thomason, Proc. Instn. Mech. Engrs., 1969/70, 184, p.896).

plastic strain ϵ_1 in a material element passing through the region in which arrowhead cracking is initiated.

In applying the basic ductile-fracture model of microvoid nucleation, growth and coalescence (Chapters 2 and 3) to the problem of establishing ductility limits in an extrusion process it is necessary to consider first the particular type of material and the required integrity of the finished product under service conditions. For example, an engineering product which is to be subjected to fatigue loading is likely to be unsuitable for service conditions if the microvoid-nucleation strain has been substantially exceeded during the manufacturing process, thus producing abundant microvoids and fatigue-crack nuclei at the site of damaged microstructural particles [11,12]. Hence, whenever fatigue resistance is required in an extruded or forged component, microvoid nucleation will represent the ductility limit in the particular metalworking process employed. A microvoid-nucleation limit of ductility will also apply generally, whatever the subsequent service conditions, to components formed from materials of relatively high particle volume fraction where the critical void-growth strains $\epsilon_1^G = (\epsilon_1^F - \epsilon_1^n)$ to initiate microvoid coalescence are always likely to be very small in comparison to the nucleation strains, (cf. Chapters

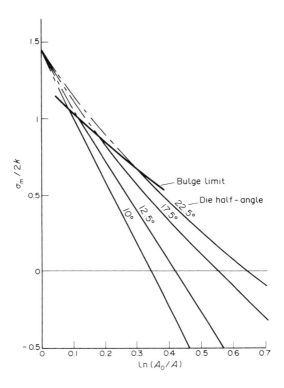

Fig. 7.8. The mean-normal stress at the axis of symmetry, in free extrusion,
for various logarithmic reductions in area and die half-angles (After
P.F. Thomason, Proc. Instn. Mech. Engrs., 1969/70, 184, p.896).

5.3 and 6.4). On the other hand, with materials of relatively low particle volume-fraction, where the void-growth strains to incipient ductile fracture can be up to an order of magnitude greater than the microvoid-nucleation strain (cf. Fig. 5.9), the ductility limit may be taken closer to the point of incipient microvoid-coalescence and ductile fracture. An illustration of the ductility limits for both the nucleation and coalescence conditions outlined above can be obtained for the four SAE steels studied in the experimental work of Le Roy et al [13] and examined in Chapter 6. With the aid of the microstructural and mechanical-property data presented by Le Roy et al [13], the Goods and Brown equation (6.14) can be evaluated to gives the microvoid-nucleation limits for the SAE steels as shown in Fig. 7.9(a); the microvoid-coalescence limits, obtained by adding the appropriate void-growth strains (from Fig. 5.9) to the nucleation strains, are shown in Fig. 7.9(b).

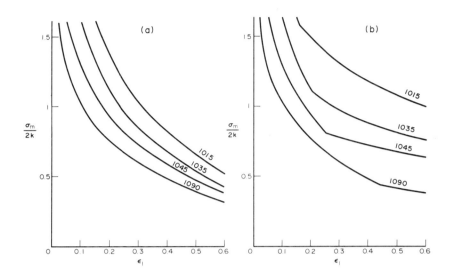

Fig. 7.9. Theoretical estimates of (a) the microvoid-nucleation strains and (b) the total ductile-fracture strains, at various mean-normal stress levels, for spheroidised SAE 1015 to 1090 steels.

The conditions under which the free-extrusion process could be carried out on the four SAE steels, without introducing microvoid-nucleation damage on the central axis of the extruded product, can now be estimated [9] by superimposing the results from Figs. 7.8 and 7.9(a), as shown in Fig. 7.10(a); this graph indicates that whenever the $\sigma_m/2k$ curves lie above the microvoid-nucleation curves, damage by microvoid-nucleation should readily occur. The results in Fig. 7.10(a) therefore suggest that only the SAE 1090 steel would be likely to exhibit gross microvoid-nucleation damage, over the intermediate range of logarithmic reductions $\ln(A_o/A)$ from 0.1 to 0.38, and this could be eliminated by using a die of 10° half-angle. On superimposing the microvoid-coalescence limits from Fig. 7.9(b) on the results from Fig. 7.8 we obtain Fig. 7.10(b), and this gives similar results to Fig. 7.10(a) with the implication that microvoid nucleation in the SAE 1090 steel would be followed almost immediately by incipient ductile-fracture and arrowhead cracking due to the very low void-growth strains required to cause microvoid coalescence at these high mean-normal stress levels.

The results in Fig. 7.8 for the mean-normal stress on the axis of an extruded bar can be modified [9] to give the equivalent results for wire drawing with zero back-tension and these are shown in Fig. 7.11, where a die half-angle of 17.5° would normally be regarded as an upper limit to the die angle in this type of process.

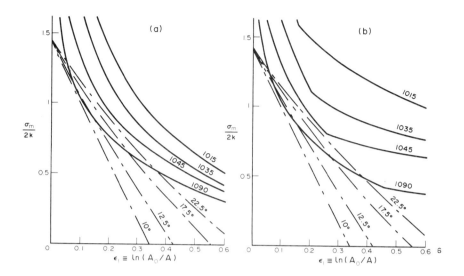

Fig. 7.10. The ductility limits in free extrusion based on (a) microvoid nucleation and (b) 'arrowhead' ductile-fracture by microvoid coalescence, from the superposition of results in Figs. 7.8 and 7.9.

Superimposing the nucleation limits in Fig. 7.11(a) suggests that wire-drawing processes will readily produce internal microstructural damage in the four SAE steels
unless the die half-angle is small; for a die half-angle of 10° or less only the SAE
1090 steel is likely to exhibit gross microstructural damage at logarithmic reductions in area below $\sim$ 0.275. On superimposing the void-coalescence limits in Fig.
7.11(b) the results suggest that arrowhead cracking should not occur at all in the
SAE 1015 steel, and only in the intermediate-reduction ranges for the SAE 1035
steel with 17.5° dies and SAE 1045 steel with 12.5° dies. These results are in good
qualitative agreement with the experimental results of Orbegozo [14] for fracture
in wire-drawing processes which show that, for a given die angle, internal
arrowhead cracking occurs only over an intermediate range of wire reduction
ratios, when the die half-angle is of the order of 15°, and the cracking can be eliminated completely by reducing the die angle.

The hydrostatic-extrusion process has a similar geometry to the free-extrusion
process, Fig. 7.6, and is performed by the application of a large fluid-hydrostatic
pressure to the extrusion billet. Early experiments [15,16] showed that with fluid-
to-air extrusion the extruded product often exhibited cracking and this could only
be eliminated by the application of a substantial fluid back-pressure P_B to the
extruded product. Without a fluid back-pressure the mean-normal stress σ_m on the

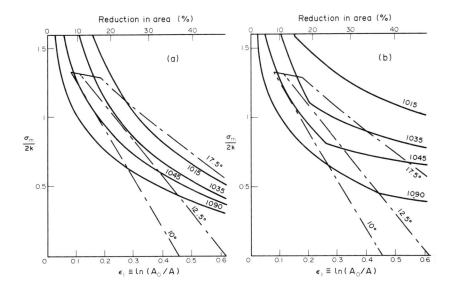

Fig. 7.11. The ductility limits in wire drawing based on (a) microvoid nucleation and (b) 'arrowhead' ductile-fracture by microvoid coalescence.

axis of symmetry, at the exit slip-line of the plastic zone, is approximately equal to that in the free-extrusion process shown in Fig. 7.8; thus the fluid-to-air hydrostatic extrusion process will have little advantage over the free-extrusion process in terms of the ductility limits. With the application of a fluid back-pressure, however, a substantial advantage is achieved by the hydrostatic process over the conventional process and this is illustrated by the results in Fig. 7.12 for hydrostatic extrusion with a back pressure of $P_B/2k = 0.5$. When the microvoid-nucleation limits for the SAE steels, from Fig. 7.9(a), are superimposed on Fig. 7.12 it appears likely that all these steels could be readily extruded against this back pressure without any apparent risk of microstructural damage.

The methods of establishing the ductility in extrusion and drawing processes described above apply in principle to all metalworking processes. However, in practical applications of the general method it would of course be necessary to take into account the effects of die friction; the results in Figs. 7.8 to 7.12 for $\sigma_m/2k$ were based on a solution with zero die-face friction. It should also be noted that the results described above do not neglect work-hardening effects since when these effects occur the yield shear stress k, for a particular value of $\sigma_m/2k$, represents the current work-hardening value of k at a strain $\epsilon_1 = \ln(A_o/A)$; and in addition the ductility

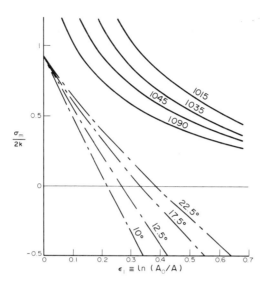

Fig. 7.12. The mean-normal stress on the axis of symmetry of the hydrostatic extrusion process, with a fluid back-pressure $P_B/2k = 0.5$,
showing that the stresses are always likely to be well below the
levels needed for microvoid nucleation in the SAE steels.

limits given in Fig. 7.9 include a full allowance for work-hardening effects in the
SAE steels.

7.3   Ductility in Sheet-Forming Processes

Industrial metal-forming processes on thin-sheet materials generally involve
drawing and stretching operations with press tools [17] under conditions
approximating to plane-stress plasticity (ie. $\sigma_3 = 0$). Hence, in order to obtain an
adequate understanding of plastic flow and ductile fracture in sheet-forming processes, it is necessary to introduce the basic equations of generalised plane-stress
plasticity [10,18,19] in which the uniform sheet thickness h is taken to be vanish-
ingly small in relation to the in-plane dimensions of the sheet and all stresses and
velocities are averaged through the thickness. (In the following presentation of
sheet-forming plasticity we shall neglect the practically-important effects of plastic
anisotropy, which can develop to a significant extent in sheet materials that have
been rolled to very large reductions in thickness).

For the case of plane-stress plasticity ($\sigma_3 = 0$) the von Mises yield criterion (1.16)

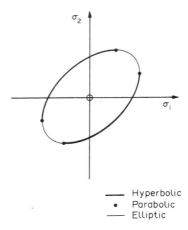

Fig. 7.13. The von Mises yield locus for plane stress states, $\sigma_3 = 0$.

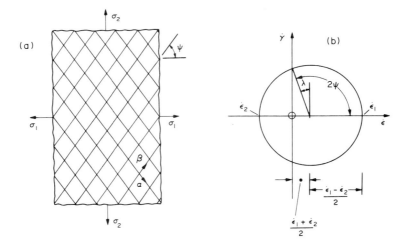

Fig. 7.14. (a) The stress- and velocity-field characteristics for a uniform
            state of generalised plane-stress plasticity.
       (b) The Mohr strain-rate circle for a plane-stress plastic velocity
            field, showing the characteristic condition and defining the
            angle λ.

reduces to the form:-

$$\sigma_1^2 - \sigma_1 \sigma_2 + \sigma_2^2 = Y^2 , \qquad (7.5)$$

and this is represented by an ellipse in the two-dimensional principal stress plane (σ_1, σ_2), Fig. 7.13. We now need to establish the conditions under which the velocity-field characteristics (α, β) will exist in a general plane-stress field, as shown in Fig. 7.14(a); where it should be noted that the velocity-field characteristics will coincide with the stress-field characteristics for a material obeying the von Mises yield criterion and associated flow rule [10]. Assuming that the material in the plane-stress field (σ_1, σ_2) obeys the Levy-Mises flow rule, the in-plane principal strain rates $(\dot{\epsilon}_1, \dot{\epsilon}_2)$ are related to the stresses by the equation:-

$$\frac{\dot{\epsilon}_2}{\dot{\epsilon}_1} = \frac{2\sigma_2 - \sigma_1}{2\sigma_1 - \sigma_2} , \qquad (7.6)$$

and thus for any given stress state (σ_1, σ_2) the Mohr strain-rate circle, Fig. 7.14(b) can be constructed. The characteristic condition of zero normal strain-rate in the direction of a characteristic defines an angle λ, which is related to the angle of the characteristics ψ by the equation $\psi = (\pi/4 + \lambda/2)$, Fig. 7.14, and from the geometry of the Mohr strain-rate circle λ has the following relations to the stress and strain-rate fields:-

$$\sin \lambda = \frac{\dot{\epsilon}_1 + \dot{\epsilon}_2}{\dot{\epsilon}_1 - \dot{\epsilon}_2} = \frac{1}{3}\left(\frac{\sigma_1 + \sigma_2}{\sigma_1 - \sigma_2}\right) . \qquad (7.7)$$

It is clear from this equation and the strain-rate circle that the characteristics only exist when :- (i) $\dot{\epsilon}_1 > 0 > \dot{\epsilon}_2$ or $\dot{\epsilon}_2 > 0 > \dot{\epsilon}_1$ and the equations are hyperbolic; (ii) $\dot{\epsilon}_1 = 0$ or $\dot{\epsilon}_2 = 0$ and the equations are parabolic [10,18]. Outside these ranges the equations are elliptic and the characteristics do not exist; the regions of the von Mises yield locus corresponding to hyperbolic, parabolic and elliptic conditions are indicated in Fig. 7.13.

The incompressibility equation (1.1) and the Levy-Mises flow rule (cf. Chapter 3.2) together establish that when $(\sigma_1 + \sigma_2) > 0$, the through-thickness strain rate $\dot{\epsilon}_3$ will be negative and all corresponding plastic states will involve sheet-thinning. On the other hand, when $(\sigma_1 + \sigma_2) < 0$ the through-thickness strain rate $\dot{\epsilon}_3 > 0$ and the corresponding plastic states involve sheet-thickening. Whenever sheet-thickening occurs the mode of failure will generally consist of plastic buckling or folding and this is not relevant to the ductile fracture problem. We therefore restrict attention in the remainder of this section to plane stress states in which $(\sigma_1 + \sigma_2) > 0$ and where sheet-thinning modes of flow are terminated by plastic necking and ductile fracture.

For plane-stress states in which the velocity-field characteristics exist, tensile plastic

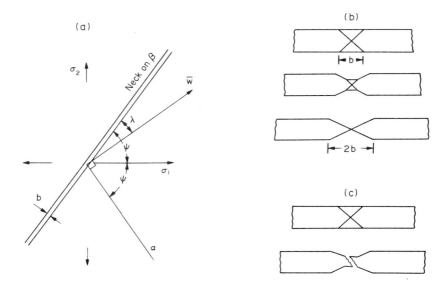

Fig. 7.15. (a) The development of a velocity discontinuity on a characteristic, in the form of a localised neck. Also showing (b) rupture
and (c) ductile fracture intervening in the necking process.

flow (with $\sigma_1 > \sigma_2$) can be terminated by the development of localised necking
along a characteristic; as shown in Fig. 7.15(a) where the relative velocity vector $\bar{w}$
across the neck is inclined at an angle λ to the neck direction and lies normal to the
direction of the other characteristic. Once a localised neck has initiated, subsequent plastic flow will be rapidly interrupted by either a full through-thickness
necking in highly ductile materials of low particle volume fraction, Fig. 7.15(b), or
by a truncated through-thickness necking when ductile fracture intervenes in materials of lower ductility and higher particle volume-fraction, Fig. 7.15(c). Clearly, the
sheet-forming process has limits of ductility established primarily by the onset of
localised necking rather than ductile fracture since even full through-thickness
necking would allow only small additional overall deformations of the sheet, once
the localised neck has begun to form. The condition under which incipient localised
necking can develop into a deep localised neck along a velocity characteristic, in
work-hardening materials, has been shown by Hill [19] to have the form:-

$$\frac{1}{Y}\frac{dY}{d\bar{\epsilon}} \le \frac{\sigma_1 + \sigma_2}{2Y} \; ; \qquad\qquad (7.8)$$

where $Y(\bar{\epsilon})$ is the uniaxial stress/strain relation and $\bar{\epsilon}$ is the equivalent strain [10]
defined by:-

$$\bar{\epsilon} = \int d\bar{\epsilon} = \sqrt{\frac{2}{3}} \int (d\epsilon_1^2 + d\epsilon_2^2 + d\epsilon_3^2)^{1/2} . \qquad (7.9)$$

Many sheet-forming processes involve plastic flow under elliptic conditions, Fig. 7.13, where velocity characteristics do not exist, and in this case a condition of diffuse necking [19], similar to that which brings about necking in cylindrical tension specimens (cf. Chapter 6.1), will usually precede ductile fracture. Diffuse necking is not related to the existence of characteristics and can occur throughout the plane-stress range of hyperbolic, parabolic and elliptic conditions; in the hyperbolic region diffuse necking usually develops at a substantially earlier stage than localised necking [19]. The diffuse-necking condition for a work-hardening material in a state of plane-stress plasticity has been shown by Swift [20] and Hill [19] to have the approximate form:-

$$\frac{1}{Y}\frac{dY}{d\bar{\epsilon}} \leqslant \frac{(\sigma_1 + \sigma_2)(4\sigma_1^2 - 7\sigma_1\sigma_2 + 4\sigma_2^2)}{4 Y^3} . \qquad (7.10)$$

It is now possible to use the localised and diffuse necking conditions to establish a forming-limit diagram for sheet-forming processes, in the restricted case where the plastic strain-rates remain in constant ratio throughout the forming process and the principal axes of the strain rates remain fixed relative to an element. Under these restricted conditions the expression (7.9) for the total equivalent strain $\bar{\epsilon}$ can be integrated to give

$$\bar{\epsilon} = \frac{2}{\sqrt{3}} \epsilon_1 \sqrt{1 + r + r^2} ; \qquad (7.11)$$

where $r = \dot{\epsilon}_2/\dot{\epsilon}_1$ and the incompressibility relation (1.1) has been used to eliminate the through-thickness strain rate $\dot{\epsilon}_3$. (NOTE: the r ratio defined above is not to be confused with the anisotropic r value which is often used as an index of the deep-drawing quality of sheet materials). If the sheet material is now assumed to work harden by a power-hardening law of the form:-

$$Y = C\,\bar{\epsilon}^m , \qquad (7.12)$$

then equations (7.7) to (7.11) give the critical strain for diffuse necking ϵ_1^D in the form:-

$$\epsilon_1^D \geqslant \frac{2m}{(1 + r)}\frac{(1 + r + r^2)}{(2 - r + 2r^2)} , \qquad (7.13)$$

and the critical strain for localised necking ϵ_1^L in the form:-

$$\epsilon_1^L \geqslant \frac{m}{(1 + r)} ; \qquad (7.14)$$

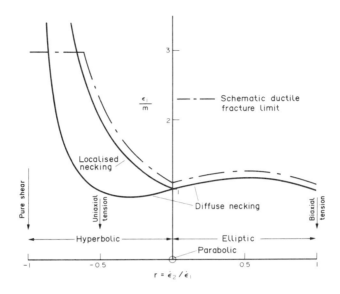

Fig. 7.16. The theoretical forming-limit diagram for plane-stress forming
        processes, based on conditions of localised necking in the
        hyperbolic region and diffuse necking in the elliptic region.

where m is the work-hardening exponent.

The results for the equalities in (7.13) and (7.14) are plotted in Fig. 7.16 in the form of the normalised principal strain ϵ_1/m against the in-plane strain-rate ratio $r = \dot{\epsilon}_2/\dot{\epsilon}_1$ in the hyperbolic ($-1 \leqslant r < 0$), parabolic ($r = 0$) and elliptic ($0 < r \leqslant 1$) regimes typically encountered in industrial sheet-forming processes; and corresponding values of σ_2/σ_1 from equation (7.6) and the characteristic angles $\psi = (\pi/4 + \lambda/2)$ from equation (7.7), for the same range of strain-rate ratios, are shown in Fig. 7.17. Over most of the hyperbolic range localised necking forms a realistic forming limit, with ductile fracture following rapidly after the inception of a localised neck, Fig. 7.15(b) and (c). However, close to the pure-shear state $r = -1$ neither diffuse nor localised necking can readily occur and in this case ductile fracture by microvoid nucleation, growth and coalescence (cf. Chapter 3) will represent the forming limit. The exact location of a ductile-fracture forming limit, that is not preceded by necking, will depend on the microstructural particle types and volume fraction in a particular material; hence this type of forming limit is represented schematically by the horizontal boundary in Fig. 7.16. It must be emphasised, however, that most practical sheet-forming operations will be carried out at r values well in excess of those that are likely to bring about ductile fracture prior to localised necking in the hyperbolic

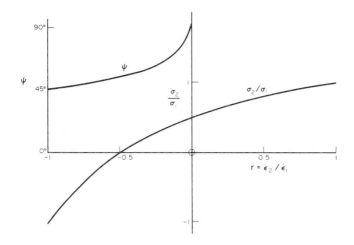

Fig. 7.17. The relationship between the stress and strain-rate ratios in plane-stress forming processes and the corresponding angular orientation ψ of the characteristics in the hyperbolic region.

range. Over the elliptic range, where localised necking cannot occur, diffuse necking represents a realistic forming limit since excessive thinning or ductile fracture is likely to follow fairly rapidly after the diffuse-necking condition has been reached. The theoretical forming-limit diagram represented by localised necking in the hyperbolic region and diffuse necking in the elliptic region, Fig. 7.16, is in good qualitative agreement with practical forming-limit diagrams obtained by Keeler and Backofen [21] and Goodwin [22] for stretching and deep-drawing processes on sheet-metal specimens, which contained printed grid-patterns for the measurement of the plastic strains.

The plane-stress sheet deformation process with localised necking is of fundamental interest in the theory of ductile fracture since it is a frequent source of experimental observations of the so called intense-shear-band mechanism of ductile fracture which has been proposed unsuccessfully as a basic mechanism of ductile fracture (cf. Chapter 3.2 and 3.5). Intense shear-bands of the type that lead to the development of ductile-fracture surfaces can only develop under conditions of relatively low geometrical constraint, where the tangential component of the velocity discontinuity can greatly exceed the normal component without violating compatibility conditions (cf. Chapter 3.2). Such kinematically-admissible velocity discontinuties or shear bands can readily develop in plane-stress modes of deformation or in the unfractured outer annular ligament of the partially-fractured neck of a uniaxial tension specimen (cf. Chapter 6.6). The most simple geometrical case

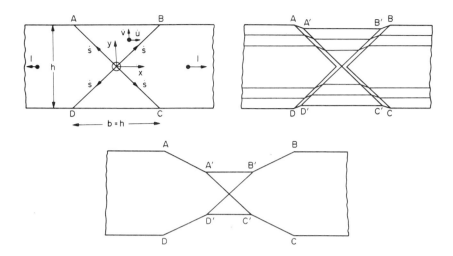

Fig. 7.18. The rigid/plastic velocity-field for a non-hardening material undergoing localised necking in plane-strain tension, showing the development of intense shear-bands at the tangential veloc-ity-discontinuities.

of localised necking along a velocity-field characteristic occurs under parabolic conditions when the neck direction coincides with the direction of the intermediate principal stress $\sigma_2 = \sigma_1/2$ and $\psi = 90°$, see Fig. 7.17. The localised neck is then subjected to pure plane-strain conditions ($\epsilon_2 = 0$) and the Onat and Prager solution [23] for the localised necking of a rigid/plastic non-hardening material represents a useful approximation. The velocity field from the Onat and Prager solution has the form shown in Fig. 7.18(a) where within the upper plastic zone AOB:-

$$\dot{u} = 0,$$
$$\dot{v} = -1,$$

(7.15)

and a tangential velocity-discontinuity $\dot{s}$ exists on the boundaries OA and OB with the form:-

$$\dot{s} = \sqrt{2}.$$

(7.16)

The velocity field in the lower plastic zone COD is a mirror-image reflection of equations (7.15) and (7.16). For a small increment of deformation in an incipient neck, with this type of velocity field, intense shear-bands are formed at the sites of the tangential velocity-discontinuity $\dot{s}$, Fig. 7.18(b), and these remain in existense throughout the necking process. Under the more general hyperbolic plane-stress conditions, where the relative velocity vector $\bar{w}$ is inclined at an angle $\lambda < \pi/2$ to the neck characteristic, Fig. 7.15(a), the incipient neck will consist of a velocity field similar to that in Fig. 7.18(a) for the parabolic case but with a superimposed shear strain-rate of $\sim \bar{w} \cos \lambda/b$ along the neck direction.

Shear bands similar to those shown in Fig. 7.18 are frequently observed in experiments on thin-sheet materials and other plastic flow fields of relatively low geometrical constraint; typical experimental results have been presented by Carlson and Bird [24] for the plane-strain necking of ferrite-austenite sheet under conditions which approximate to a parabolic state ($\dot{\epsilon}_2 = 0$). Intense shear bands therefore do not in themselves represent a basic mechanism of ductile fracture and actually represent a tangential discontinuity in the incompressible-plastic velocity field, under conditions of relatively low geometrical constraint, where the basic mechanisms of nucleation, growth and coalescence of microvoids must develop within the regions of intense shear-strain to bring about localised ductile fracture (cf. Chapters 6.6 and 8.2).

7.4   Thin-Sheet Models of Ductile Fracture.

Following the presentation of the equations of plane-stress plasticity in the previous section it is interesting at this point to consider the validity of a plane-stress model of ductile fracture which has been presented recently [25]. A regular two-dimensional array of circular holes in a thin-sheet specimen can be represented by one of the two plane-stress characteristic fields [19] shown in Fig. 7.19, depending on the ratio of hole spacing to hole diameter e/a. Now the constraint factor σ_n/Y for plane-stress plastic deformation between two holes of radius a and spacing 2e. in a thin sheet, has been shown by Hill [19] to have the form:-

$$\frac{\sigma_n}{Y} \approx 1 + 0.226 \, \frac{e}{e+a} \,, \quad 0 \leqslant \frac{e}{a} \leqslant 1.071 \qquad (7.17a)$$

$$\frac{\sigma_n}{Y} = \frac{2}{\sqrt{3}} - 0.040 \, \frac{a}{e} \,, \quad \frac{e}{a} \geqslant 1.071 \,. \qquad (7.17b)$$

The uniform stress σ_1^{1c} on the ends of a plane-stress sheet specimen (Fig. 7.19), corresponding to these values of σ_n, are therefore given by:-

$$\frac{\sigma_1^{1c}}{Y} = \left(\frac{e}{e+a}\right)\left(1 + 0.226 \, \frac{e}{e+a}\right), \quad 0 \leqslant \frac{e}{a} \leqslant 1.071 \; ; \qquad (7.18a)$$

$$\frac{\sigma_1^{1c}}{Y} = \left(\frac{e}{e+a}\right)\left(\frac{2}{\sqrt{3}} - 0.040 \, \frac{a}{e}\right), \quad \frac{e}{a} \geqslant 1.071 \,. \qquad (7.18b)$$

It is clear from these equations that the constraint factors for plane-stress conditions

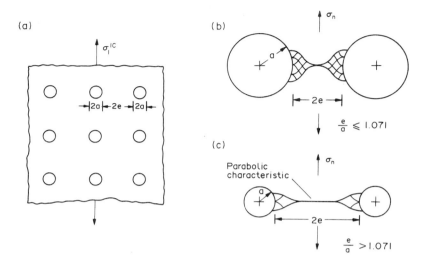

Fig. 7.19. (a) A thin-sheet specimen with a regular array of holes. Also
showing the characteristic fields for plane-stress plastic linking
of the holes when (b) the holes are closely spaced and (c) the
holes are wide apart.

are very much lower than those encountered under the analogous plane-strain (cf.
Figs. 3.5 and 3.13) and three- dimensional (cf. Fig. 5.7) conditions, and this is illustrated in Fig. 7.20 by comparing the plane-stress results from equation (7.18b) with
the plane-strain results for circular holes from Fig. 3.13. These results show that
plane-stress models are virtually incapable of modelling the void-growth stage of
ductile fracture because when $e/a < 6.72$ the applied stress σ_1^{1c} is less than the uniaxial yield stress Y, and thus all plastic deformation will be concentrated in the inter-
hole ligaments. This critical ratio $e/a \sim 6.72$ represents a hole volume-fraction $V_f \sim$
0.013, hence, whenever the hole volume-fraction in a plane-stress sheet specimen
exceeds this value (ie. $V_f > 0.013$) plastic deformation will become localised at a
single row of holes almost immediately after the first small increment of plastic
deformation. Of course, in annealed materials, which initially display a strong
work-hardening effect, the immediate tendency to develop localised flow between
the holes may be delayed slightly.

When $e/a > 6.72$ the plastic constraint factor under plane-stress conditions is still
very small giving a maximum value $\sigma_1^{1c}/Y \to 1.155$ as $e/a \to \infty$ (Fig. 7.20), which is
at least an order of magnitude lower than the values of $\sigma_1^{1c}/2k$ from the equivalent
plane-strain model. Hence, even with hole volume-fractions $V_f \ll 0.013$ only very
small void-growth strains would be needed to initiate localised plane-stress neck-

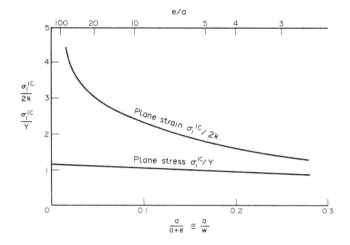

Fig. 7.20. The effect of plastic constraint on the stresses needed to bring about hole-linking by plastic limit-load failure of the inter-hole material, for both plane-stress and plane-strain conditions.

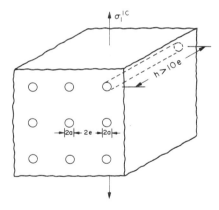

Fig. 7.21. The required geometry to give a valid plane-strain model of ductile void-growth and coalescence.

ing at a row of holes, thus giving a completely inadequate model of the ductile-fracture process. In fact the hole volume-fraction used in the plane-stress model of ductile fracture [25] is $V_f \sim 0.05$ and there is clearly no possibility of simulating the microvoid-growth process with such high values of V_f. A further important consequence of the critical hole volume-fraction $V_f \sim 0.013$, above which plane-stress plastic deformation is almost immediately localised at a single row of holes, is that the plane-stress conditions for 'void coalescence' are virtually independent of hole volume-fraction; this is completely at variance with experimental results which show that ductile-fracture strains are strongly dependent on the volume-fraction of microvoids (cf. Chapters 3 and 5). There is therefore no theoretical basis for the validity of plane-stress models of ductile fracture, although the results might be of some value in studies involving the linking of pre-existing defects in sheet-forming processes.

An experimental model of the ductile-fracture process in metals can be designed, but this requires a plane-strain rather than a plane-stress model. The plane-strain 'void coalescence' condition could be achieved in a specimen of rectangular cross-section where the through-thickness dimension h is at least five times greater than the hole spacing 2e, ie. $h > 10e$ (Fig. 7.21). With this type of model very large plastic constraint factors are possible when e/a is large and this will require large hole-growth strains to bring about 'void coalescence' by internal necking , analogous to the effects occuring in real ductile-fracture processes (cf. Chapter 3).

REFERENCES

1. Thomason, P.F., *Proc. Instn. Mech. Engrs.,* 1969/70, 184, p.875.
2. Thomason, P.F., *Int. J. Mech. Sci.,* 1969, 11, p.65.
3. Billigmann, J., *Draht,* 1950, 1, p.49.
4. Billigmann, J., *Stauchen und Pressen,* Bild 46, Carl Hanser Verlag, München, 1953.
5. Jenner, A., and Dodd, B., *J. Mech. Work. Tech.,* 1981, 5, p.31.
6. Bradbury, T.G., *Proc. Met. Soc. Conf. Pittsburgh,* 1961, 12, p.29.
7. Sachs, K., and Bestwick, R.D.W., *Proc. Instn. Mech. Engrs.,* 1969/70, 184, p.906.
8. Thomason, P.F., *Proc. Instn. Mech. Engrs.,* 1969/70, 184, p.885.
9. Thomason, P.F., *Proc. Instn. Mech. Engrs.,* 1969/70, 184, p.896.
10. Hill, R., *The Mathematical Theory of Plasticity,* Clarendon Press, Oxford, 1950.
11. Eid, N.M.A., and Thomason, P.F., *Acta Metall.,* 1979, 27, p.1239.
12. Eid, N.M.A., and Thomason, P.F., *Analytical and Experimental Fracture Mechanics,* (Edited by G.C. Sih and M. Mirabile), Sijthoff and Noordhoff, The Netherlands, 1981.
13. Le Roy, G., Embury, J.D., Edwards, G., and Ashby, M.F., *Acta Metall,* 1981, 29, p.1509.
14. Orbegozo, J.I., *Annals C.I.R.P.,* 1968, 16, p.319.
15. Pugh, H.Ll.D., *Int. Res. Prod. Eng.,* 1963, A.S.M.E., p.394.
16. Fiorentino, R.J., Richardson, B.D., and Sabroff, A.M., *Metal Forming,* September 1969, p.243.

17. Alexander, J.M., *Metall. Rev.,* 1960, 5, p.349.
18. Kachanov, L.M., *Fundamentals of the Theory of Plasticity,* MIR Publishers, Moscow, 1974.
19. Hill, R., *J. Mech. Phys. Solids,* 1952, 1, p.19.
20. Swift, H.W., *Brit. Iron & Steel Res. Ass. Rep.,* 1950, MW/E/S2/50,
21. Keeler, S.P., and Backofen, W.A., *Trans A.S.M.,* 1963, 56, p.25.
22. Goodwin, G.M., *S.A.E. Automotive Eng. Cong.,* Detroit, 1968, paper 680093.
23. Onat, E.T., and Prager, W., *J. Appl. Phys.,* 1954, 25, p.491.
24. Carlson, J.M., and Bird, J.E., *Acta Metall.,* 1987, 35, p.1675.
25. Magnusen, P.E., Dubensky, E.M. and Koss, D.A., *Acta Metall.,* 1988, 36, p.1503.

CHAPTER 8

Ductile Fracture in Notched Bars
and Cracked Plates

8.1 Introduction

When ductile fracture occurs in the plastic zones at the roots of notches and sharp cracks, discontinuities in the local plastic velocity-fields can have a strong influence on the way in which ductile-fracture initiates. For example, the incompressible plastic velocity-fields for notched bars in tension and bending often contain tangential velocity-discontinuities and these become the sites for the development of bands of intense shear-strain where the ductile fracture mechanisms of microvoid nucleation, growth and coalescence develop highly locally within the shear bands (cf. Section 8.2). In contrast, when the plane-strain plastic zone at the tip of a sharp crack is relatively small and totally enclosed by an elastic/plastic boundary, tangential velocity discontinuities are kinematically inadmissible and bands of intense shear-strain cannot develop (cf. Section 8.3). Hence, under these conditions, the ductile-fracture mechanisms of microvoid nucleation, growth and coalescence operate in a 'continuous' plastic strain-rate field where it is possible to develop quantitative models of the ductile-fracture processes, based on plane-strain slip-line field theory.

A somewhat different approach is required for problems involving ductile fracture at the root of cracks in thin-sheet materials. In this case the crack-tip plastic field is almost entirely subjected to plane-stress plasticity and it becomes necessary to consider states of hyperbolic and parabolic plasticity and the conditions for localised necking along velocity-field characteristics (cf. Section 8.4). The final parts of this Chapter are then devoted to problems of elastic-plastic fracture mechanics where ductile fracture brings about substantial amounts of stable crack-growth prior to catastrophic fracture (cf. Section 8.5).

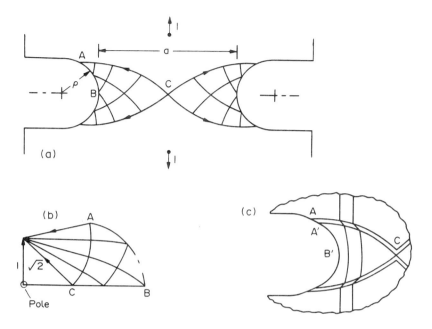

Fig. 8.1. (a) A notched bar in plane-strain tension showing (b) the hodograph with tangential velocity-discontinuities and (c) the localised
shear-bands on the bounding slip-lines where the velocity discontinuities exist.

8.2 Ductile Fracture at the Root of Blunt Notches, under Tension and Bending.

Experimental studies of ductile fracture at the root of blunt notches can give mis-
leading results unless it is recognised that intense shear-deformation bands can
develop in an incompressible-plastic velocity field as a natural feature of kinematically admissible plastic deformation under plane-strain conditions [1,2]. The
bands of intense shear strain in plane-strain plastic deformation zones are located
at the sites of tangential velocity-discontinuities in the plastic velocity field of the
equivalent rigid/plastic non-hardening solid, and although strong work-hardening
effects in real materials will tend to increase the thickness of the shear bands and
reduce somewhat the intensity of the localised shear strains they nevertheless
remain as bands of highly-elevated shear strain.

A typical example of a notch-root plastic field, with tangential velocity-discontinuities
on the bounding slip-lines of the plastic deformation zone, is shown in Fig. 8.1(a) for
a deeply-notched bar with semicircular notch roots, pulled in axial tension [1]. Using

the hodograph for the plastic velocity-field (Fig. 8.1(b)), the deformation of a set of axial grid lines, corresponding to a small axial extension of the notched bar is shown in Fig. 8.1(c). The deformed grid lines show the development of bands of intense shear strain, at the site of the tangential velocity-discontinuities, which will clearly lead to a localisation of the microvoid nucleation, growth and coalescence mechanisms of ductile fracture, within these regions, for real materials; such effects have been observed in experiments [3]. It is important to note, however, that these intense shear-bands are not directly related to the 'shear-banding' effect in dilational-plastic solids, as proposed in the Berg-Gurson theory of ductile fracture (cf. Chapter 3.5). The proposed shear-banding effect in the Berg-Gurson model can only develop *after* the nucleation of microvoids; in contrast the intense shear-bands discussed above can readily develop in incompressible-plastic fields *before* microvoids have been nucleated. Hence, the experimental observation of ductile fracture along intense shear-bands does not give support to the Berg-Gurson model of ductile fracture.

Whenever the plane-strain plastic field at a notch root contains tangential velocity-discontinuities, the ductile fracture process is likely to be confined to the bands of intense shear-strain and will be difficult to model on a quantitative basis; one of the main problems being the difficulty of modelling the microvoid-growth process when the principal axes of the microvoids rotate relative to the principal axes of successive strain-increments. However, it may be possible in these problems to regard microvoid nucleation as the critical event controlling ductile fracture in an intense shear-band, since the external deformation-increment between microvoid nucleation and coalescence is likely to be small.

Notched bars in pure bending can in certain cases exhibit plane-strain plasticity with a velocity field that does not contain tangential velocity-discontinuities; thus giving a smoothly varying plastic strain-rate field without intense shear bands. An example of this type of plastic flow field occurs for the plane strain bending of a double-notched bar [4,5], with circular notch roots of radius ρ and geometry $\rho/a >$ 1.513, Fig. 8.2(a), where it was shown by Green [5] that conditions of positive plastic-work preclude the existence of velocity discontinuities on slip-lines passing through a point N of bending-stress discontinuity. On the other hand a similar type of double-notched bar, of geometry $\rho/a \leqslant 1.513$ [4,5], does contain tangential velocity-discontinuities which emerge at the points A,B,C and D on the notch surfaces, Fig. 8.2(b), thus introducing intense shear-hands into the plastic flow field similar to those shown schematically in Fig. 8.2(c). With this type of velocity field the central rigid region EF, enveloped by tangential velocity-discontinuities, acts as a plastic hinge about which rotation occurs. Similar plastic fields develop in single-notched specimens, with circular notch roots, and notch-geometry parameters of $\rho/a > 0.563$, where velocity discontinuities do not exist (Fig. 8.3(a)), and $\rho/a \leqslant 0.563$ where they do exist (Fig. 8.3(b)).

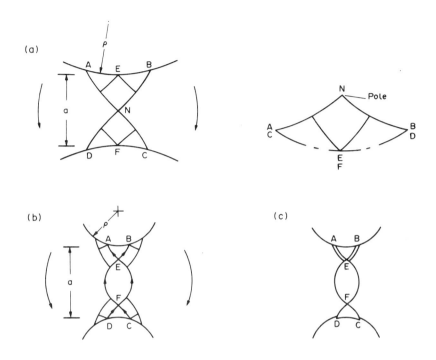

Fig. 8.2. The slip-line fields for double-notched bars in bending: (a) when $\rho/a > 1.513$ and the hodograph contains no velocity discontinuities; (b) when $\rho/a \leqslant 1.513$ and tangential velocity-discontinuities produce the localised shear-bands shown in (c).

A state of pure bending, with zero resultant shear-force across the minimum section, is achieved approximately in four-point bending. In three-point bending or cantilever bending where the minimum section is subjected to a relatively high shear-force, similar to the conditions in Charpy and Izod specimens, plane-strain plastic fields with tangential velocity-discontinuities are even more prevelant than in pure bending [6,7,8] and the notch-root plastic zone will invariably contain intense shear bands. A further complicating effect of the notch geometry in Charpy and Izod specimens, in relation to experimental investigations of notch-root ductile fracture, is that the notches are not "sufficiently deep" to ensure containment of the notch-root plastic field and thus the maximum possible plastic constraint is not achieved and the maximum stresses are significantly less than the theoretical maximum [7,8]. It is clear from the above analysis that studies of ductile fracture in

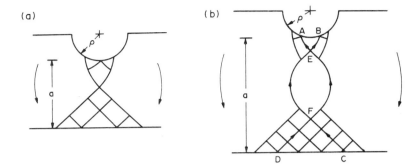

Fig. 8.3. The slip-line fields for single-notched bars in bending: (a) when
$\rho/a > 0.563$ and velocity discontinuities are not present; (b) when
$\rho/a \leqslant 0.563$ and velocity discontinuities lead to localised shear-
bands on AD and BC.

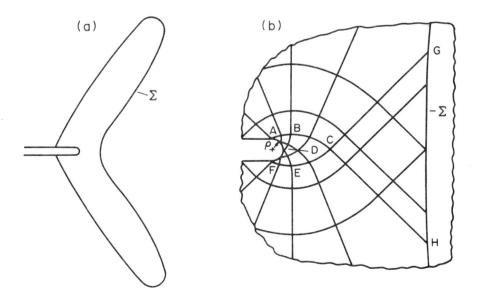

Fig. 8.4. (a) The plastic zone at the tip of a sharp crack in mode I loading.
(b) The slip-line field between the crack tip and the adjacent elas-
tic/plastic boundary.

notched bars must be made on specimens of carefully chosen geometry if discontinuities in the plastic velocity-fields are to be avoided.

8.3   Ductile Fracture at the Root of Cracks under Plane-Strain Conditions.

The plane-strain plastic zone at the tip of a crack, in a real work-hardening elastic-plastic material, has the form shown schematically in Fig. 8.4(a) for the case of mode I crack-opening loading [9,10,11]. For an equivalent plastic-rigid non-hardening model, with a circular crack-tip of radius ρ, the slip-line field in the plastic zone ACF (Fig. 8.4(b)) is uniquely determined by the zero-stress boundary conditions on the free surface ADF (Cauchy problem) [1,4]. The crack-tip zone ACF of logarithmic spiral slip-lines [1,4] can undergo large plastic straining which is accommodated in the adjacent non-centred fans extending over relatively large distances to the elastic/plastic boundary, Fig. 8.4. The zero-stress boundary conditions on the adjacent flat surfaces of the crack create uniform plastic stress fields with straight slip-lines that lead directly to a further uniform-stress field CGH ahead of the crack-tip zone ACF, and this uniform plastic field will extend to the elastic/plastic boundary. The uniform stress field CGH of straight α and β slip-lines (Fig. 8.4(b)) must clearly be a region of only elastic-order displacements and strains due to the constraints imposed by the elastic/plastic boundary and this will also have the effect of excluding the development of tangential velocity-discontinuities in the crack-tip flow field. The slip-line field and hodograph for the crack-tip zone ACF is shown in Fig. 8.5 where the pole point for the hodograph (Fig. 8.5(b)) at point C represents the virtually zero velocities in the entire elastically-constrained region CGH.

The stress distribution on the x axis of symmetry of the crack-tip zone is given in terms of the σ_y and mean-normal stress σ_m components by the equations [1,2,4]:-

$$\frac{\sigma_y}{2k} = 1 + \ln\left(1 + \frac{x}{\rho}\right),\tag{8.1a}$$

$$\frac{\sigma_m}{2k} = 0.5 + \ln\left(1 + \frac{x}{\rho}\right),\tag{8.1b}$$

where k is the yield shear stress of the non-hardening material. These equations are valid over the logarithmic-spiral region $0 \leqslant x/\rho \leqslant 3.81$, and for values of $x/\rho > 3.81$ the normalised stresses $\sigma_y/2k$ and $\sigma_m/2k$ will remain constant up to the elastic/plastic boundary, Fig. 8.6(a). Since most engineering materials will exhibit a pronounced work-hardening effect under the large plastic strains that develop in the crack-tip region, an approximate allowance for these effects can be made by using an appropriately work-hardened value of yield-shear stress k in place of the initial-yield value. A measure of the plastic strains in the crack-tip region can be made with the aid of the hodograph by considering deformation in a triangular region

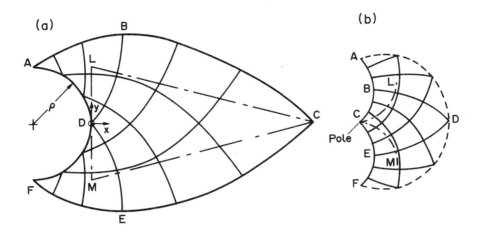

Fig. 8.5. (a) The slip-line field and (b) hodograph for the crack-tip plastic zone.

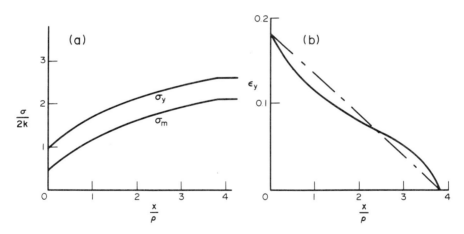

Fig. 8.6. The stress and strain distributions on the axis of symmetry of the crack-tip plastic zone.

LCM (Fig. 8.5(a)) of height 2ρ and length 3.81ρ. The velocities along the top (LC) and bottom (MC) surfaces of the triangular region are given by the images of LC and MC in the hodograph (Fig. 8.5(b)) and, on the assumption that the strain rate $\dot{\epsilon}_y$ at any point x/ρ is uniform for all values of y within the boundaries of the triangle, the strain ϵ_y across the triangular region LCM is shown in Fig. 8.6(b) for the case of vertical displacements of $\pm \rho/5$ at L and M, respectively. It is clear from Fig. 8.6(b) that the ϵ_y distribution does not depart substantially from a simple linear form and it is therefore possible to represent the ϵ_y strains in LCM by the linearised expression:-

$$\epsilon_y = \left(1 - \frac{x/\rho}{3.81}\right) \ln\left(1 + \frac{\Delta}{\rho}\right) \approx \left(1 - \frac{x/\rho}{3.81}\right) \frac{\Delta}{\rho} \; ; \tag{8.2}$$

where Δ is the magnitude of the displacements of the points L and M in the $\pm$ y direction, respectively, for a unit time-increment.

It can be seen from the hodograph, Fig. 8.5(b), that the displacements of the notch-root surface ADF will leave the notch root in an approximately circular form after large plastic flow and so equations (8.1) and (8.2) can be used to evaluate the stress and strain states in the crack-tip plastic zone for a substantial degree of crack-opening displacement. The crack-tip stress fields will thus have an unchanging form as the crack-tip radius increases, but with each increment of crack-tip radius the sum of the plastic strains ϵ_y will increase as shown schematically in Fig. 8.7 to give an increasingly non-linear distribution for large crack-opening displacements. However, for present purposes it can be assumed that the crack-opening displacement remains small enough to retain the linear form of equation (8.2).

The relative displacement 2Δ of the points L and M on the triangular region LCM (Fig. 8.5(a)), within the crack-tip plastic zone, can be taken as being broadly equivalent to the crack-opening displacement δ and this is given to a close approximation, for plane-strain plasticity at the crack tip, by the expression [9-12]:-

$$\delta = \frac{0.6 \, K_I^2}{Y_o \, E} \; , \tag{8.3}$$

where K_I is the stress intensity factor in mode I loading, E is Young's modulus of elasticity and Y_o is the initial uniaxial yield stress. The linearised expression for the ϵ_y strain in the crack-tip plastic zone can therefore be rewritten in the form:-

$$\epsilon_y \approx \frac{0.3 \, K_I^2}{\rho \, Y_o \, E} \left(1 - \frac{x/\rho}{3.81}\right) \; . \tag{8.4}$$

Equations (8.1b) and (8.4) can now be used to examine the conditions leading to incipient ductile fracture in the crack-tip region.

When the microstructural second-phase particles and inclusions in the crack-tip region have a radius of the order $1\mu m$ or less the conditions for microvoid nucleation are given by the Goods and Brown model [13] (cf. Chapter 2.1) and this can be written for the plane-strain plastic field in the form:-

$$\epsilon_y^n = \left(\frac{\sigma_c - \sigma_m}{H}\right)^2 \; ; \tag{8.5}$$

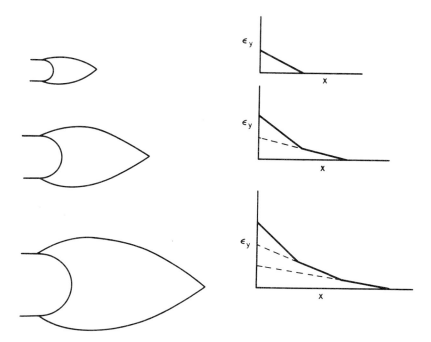

Fig. 8.7. A schematic representation of the effect of increasing Mode I
loads on the crack-tip plastic zone and the total plastic strain ϵ_y.

where σ_c is the critical cohesive strength of the matrix/particle interface and H is a
material constant related to the particle size. An estimate of the microvoid nucleation strain, at any point x/ρ in the triangular region LCM at the crack tip, can now be
obtained from equations (8.1b) and (8.5) with a suitable correction for work-hardening effects. Assuming that the crack-tip material obeys a power-hardening law
similar to equation (6.6), the mean-normal stress at any point x/ρ in the crack-tip
region is given approximately by the expression:-

$$\sigma_m = \left(\frac{2}{\sqrt{3}} \right)^{m+1} C \, \epsilon_y^m \left(0.5 + \ln \left(1 + \frac{x}{\rho} \right) \right) ; \qquad (8.6)$$

where C is a material constant and m is the work-hardening exponent. Combining
equations (8.5) and (8.6) gives the following approximate expression for estimating the microvoid-nucleation strain ϵ_y^n at x/ρ:-

$$H \left(\epsilon_y^n \right)^{0.5} = \sigma_c - \left(\frac{2}{\sqrt{3}} \right)^{m+1} C \left(\epsilon_y^n \right)^m \left(0.5 + \ln \left(1 + \frac{x}{\rho} \right) \right) . \qquad (8.7)$$

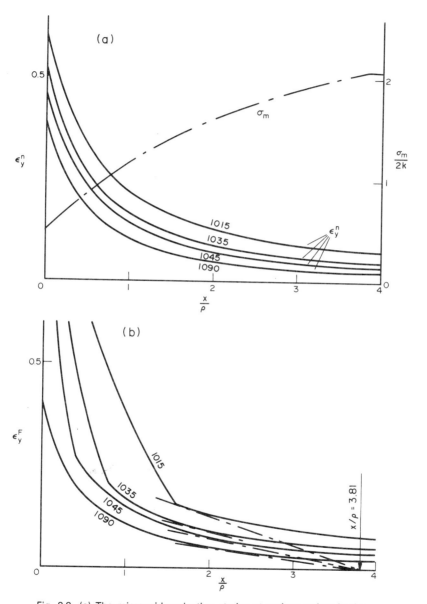

Fig. 8.8. (a) The microvoid-nucleation strains at various points in the crack-tip plastic zone for a number of SAE steels.
(b) The corresponding ductile-fracture strains for the SAE steels.

This type of equation can be solved for integer values of $(2m)^{-1}$ and when m takes the typical value for steels of $m \approx 0.25$ [14] a quadratic equation in $(\epsilon_y^n)^{0.25}$ is obtained with the solution :-

$$(\epsilon_y^n)^{0.25} = \frac{-1 \cdot 2\,C\left(0.5 + \ln\left(1 + \frac{x}{\rho}\right)\right) + \sqrt{1.43\,C^2\left(0.05 + \ln\left(1 + \frac{x}{\rho}\right)\right)^2 + 4\,H\sigma_c}}{2H}$$

(8.8)

Using values of $H = 1071\,\text{MPa}$ and $\sigma_c = 1200\,\text{MPa}$ for carbide particles in steels (cf. Chapter 6.3), and appropriate values of the work-hardening constant C for the SAE steels (Table 6.1) in the work of Le Roy et al [14], the approximate microvoid-nucleation strains at various points x/ρ in a crack-tip field are shown in Fig. 8.8(a). The total ductile-fracture strains ϵ_y^F at the points x/ρ, obtained by adding the critical void-growth strains ϵ_y^G for plastic limit-load failure of the intervoid matrix (cf. Chapters 3.3, 3.4, 5.3, 6.4 and 7.2) to the microvoid-nucleation strains are shown for the same SAE steels in Fig. 8.8(b).

The critical conditions leading to incipient ductile fracture in the crack-tip region can now be determined by superimposing the linear ϵ_y strain equation (8.4) on the fracture strain ϵ_y^F graphs in Fig. 8.8(b). For each steel in turn a continuously increasing K_I value brings about the critical condition for ductile fracture when the ϵ_y strain distribution forms a point of tangency with the microvoid-nucleation ϵ_y^n curve; ie. when $\epsilon_y = \epsilon_y^n$ and $d\epsilon_y/dx = d\epsilon_y^n/dx$. This result suggests that, except for materials with very low volume-fractions of microstructural particles, ductile fracture will initiate well ahead of the crack tip and will be nucleation-controlled with microvoid coalescence following spontaneously at the point of microvoid nucleation. The critical point of tangency for the four steels in Fig. 8.8(b) occurs at $x/\rho \approx 1.9$ and on substituting this value in equations (8.4) and (8.8) and equating ϵ_y to ϵ_y^n with $K_I \equiv K_{IC}$, we obtain the following expression for K_{IC}:-

$$K_{IC} = 2.58\sqrt{EY_o\,\rho}\left(\frac{+\sqrt{3.53C^2 + 4H\sigma_c} - 1.88C}{2H}\right)^2$$

(8.9)

which is valid for a material with a work-hardening exponent $m \approx 0.25$. Now it has been shown in experiments [15] that the effective radius of a sharp crack is of the order 0.05 mm and a typical value of E for steels is 210 GPa; hence equation (8.9) can be rewritten for steels with an exponent $m \approx 0.25$ in the form:-

$$K_{IC} = 7.38\sqrt{Y_o}\left(\frac{+\sqrt{C^2 + 1.13\,H\sigma_c} - C}{H}\right)^2.$$

(8.10)

This expression is of course only valid under conditions where microvoid nucleation controls the crack-tip fracture process, and where the microstructural particles are sufficiently small for the validity of the Goods and Brown nucleation model [13]. The material constant C in the work-hardening law and the parameter H in the microvoid nucleation model [13] are both functions of the microvoid volume fraction V_f and so K_{IC} is strongly dependent on Y_o, σ_c and V_f. The K_{IC} expression (8.10) also suggests a strong increase in fracture toughness with increased test temperature and reduced loading rate, which are likely to promote local recovery effects at the sites of microstructural particles thus making microvoid nucleation more difficult [13] (cf. Chapter 4.3). Typical values of the parameters in equation (8.10) for an SAE 1045 steel [14] are $Y_o = 302$ MPa, $C = 940$ MPa, $H = 1071$ MPa and $\sigma_c = 1200$ MPa, and on substituting these values the fracture toughness is predicted to be of the order $K_{IC} = 38.74$ MPa$\sqrt{m}$, which is about the right order for this material.

Models of ductile fracture at the tips of sharp cracks, of the order of 50μm root radius [15], are likely to be particularly sensitive to the absolute size and spacing of the various populations of microstructural particles that can exist in engineering materials (cf. Chapters 2.1 and 3.1). For example a typical quenched and tempered steel may contain a volume fraction of carbide particles of $\approx 2\mu$m diameter and a volume fraction of aluminium-oxide particles of $\approx 20\ \mu$m diameter. If the volume fraction of carbides is very much greater than that of the oxides then the carbides are likely to have the controlling influence on crack-tip fracture and equation (8.10) would represent a valid model for the fracture toughness parameter K_{IC}, Fig. 8.9(a). On the other hand if the volume-fraction of carbides is very much less than that of the oxides a larger degree of crack-tip deformation will be required to increase the crack-tip radius and enclose a sufficient number of the oxide particles, within the logarithmic spiral field of large plastic-strain and high mean-normal stress, to form a fracture surface, Fig. 8.9(b). These effects need to be taken into consideration when developing models of ductile fracture at the tips of sharp cracks.

8.4   Ductile Fracture at the Root of Cracks under Plane-Stress Conditions.

It is instructive to begin an analysis of the plane-stress crack problem with the elastic stress-field equations for a crack of small but finite root-radius ρ, Fig. 8.10. By the simple artifice of moving the origin of the polar coordinate system (r, θ) to a point $\rho/2$ behind the crack tip Creager and Paris [16] obtained stress field equations without a crack-tip singularity which reduce to the following form on the axis of symmetry $(\theta = 0)$:-

$$\sigma_x = \frac{K_I}{\sqrt{2\pi r}} \left[1 - \frac{\rho}{2r} \right] , \qquad\qquad (8.11a)$$

$$\sigma_y = \frac{K_I}{\sqrt{2\pi r}} \left[1 + \frac{\rho}{2r} \right] , \qquad\qquad (8.11b)$$

$$\tau_{xy} = \sigma_z = 0 , \qquad\qquad (8.11c)$$

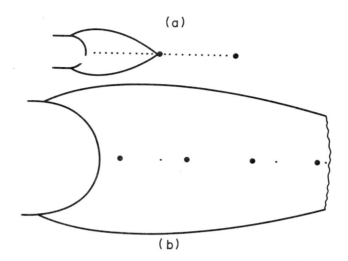

Fig. 8.9. A schematic representation of the crack-tip plastic zone at incipient ductile fracture when:- (a) the volume-fraction of carbide
particles greatly exceeds that of the oxides; (b) the volume-fraction of carbides is very much less than that of the oxides.

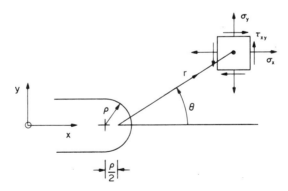

Fig. 8.10. The removal of the crack-tip singularity by locating the polar-
coordinate origin at $\rho/2$ behind the crack tip, following the
method of M. Creager and P.C. Paris [16].

where K_I is the appropriate mode I stress intensity factor. These equations can be
evaluated for various values of the normalised coordinate r/ρ to give the stress distributions ahead of the crack tip as shown in Fig. 8.11(a). It is clear from these
results that as K_I is gradually increased the initial state of plane-stress plasticity will

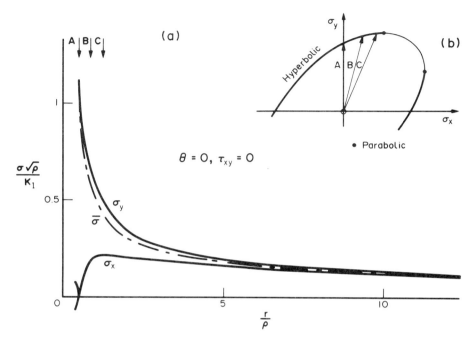

Fig. 8.11. (a) The in-plane stresses σ_x, σ_y and the equivalent stress $\bar{\sigma}$ on the axis of symmetry of a plane-stress crack-tip field.
(b) The corresponding von Mises yield locus.

begin at the crack tip, and will be well within the region of hyperbolic plasticity on the yield locus [1,4] (cf. Chapter 7.3). The plane-stress states at points (A,B,C) on the axis of symmetry ahead of the crack are represented schematically by the stress paths (A,B,C) on the von Mises yield locus (Fig. 8.11(b)) for the case of a monotonically increasing K_I value, and it is clear that a completely hyperbolic-plasticity region is likely to develop initially ahead of the crack tip [17]. This hyperbolic plane-stress plastic field will have the form shown schematically in Fig. 8.12, which is similar to the ones used by Thomason [17] and Nishimura and Achenbach [18], but differs significantly from the one proposed by Hutchinson [19] where parabolic conditions existed along the axis of symmetry of the crack-tip field.

For thin-sheet materials of relatively high ductility the ductile growth of a crack, with elastically contained plane-stress plastic zones, will be strongly influenced by the exact nature of plane-stress plasticity existing ahead of the crack tip. When the plastic strains in the crack-tip field are sufficiently large to satisfy the conditions for both diffuse necking (equation (7.12)) and localised necking (equation (7.13)) there

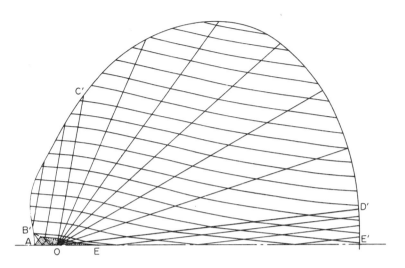

Fig. 8.12. The crack-tip plastic zone and characteristic field for completely hyperbolic plane-stress conditions. (After P.F. Thomason, "Fracture Mechanics in Engineering Application", edited by G.C. Sih and S.R. Valluri, Sijthoff and Noordhoff, The Netherlands, 1979. Reprinted by permission of Kluwer Academic Publishers).

will be a tendency for localised necking to develop along a velocity-field characteristic [4,20]. However, for hyperbolic characteristics the relative velocity vector $\bar{w}$ across a neck is inclined at an angle λ to the characteristic (Fig. 7.15); where λ is given by equation (7.7) [1,4,20]. Now it is clear from equation (7.7) that $\lambda < \pi/2$ for all hyperbolic characteristics and thus the elastic/plastic boundary will put a strong constraint on displacements $\bar{w} \cos \lambda$ along any hyperbolic characteristic at incipient necking, thus effectively preventing the development of a neck along this type of characteristic. However, if a continuously increasing K_I value were to change the plane-stress field at the crack tip from a completely hyperbolic field to one in which a parabolic characteristic developed along the axis of symmetry ahead of the crack [17] the angle λ of the relative-velocity vector $\bar{w}$ would now lie at $\lambda = \pi/2$ to the single parabolic characteristic. Thus $\bar{w} \cos(\pi/2) = 0$ and the elastic/plastic boundary can no longer have a constraining effect on the development of a deep neck along the parabolic characteristic ahead of the crack; the mechanisms of ductile fracture could then operate rapidly within the localised-necking region (Fig. 7.15) to bring about ductile crack growth [21]. This suggests that the critical conditions for ductile crack-growth under plane-stress conditions are likely to be associated with the transformation of the crack-tip plastic field from a completely hyperbolic form to a hyperbolic/parabolic form, Fig. 8.13, [17]. If this is

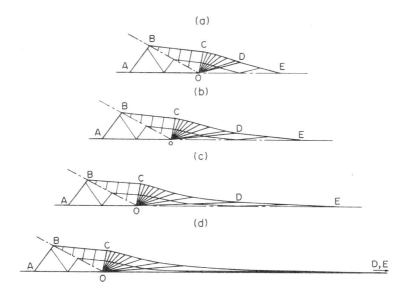

Fig. 8.13. Statically-determined characteristic fields for various orientations of the stress discontinuity BO giving completely hyperbolic plane-stress plasticity (a) - (c), but hyperbolic/parabolic
plasticity (d). (After P.F. Thomason, "Fracture Mechanics in
Engineering Application", edited by G.C. Sih and S.R. Valluri,
Sijthoff and Noordhoff, The Netherlands, 1979. Reprinted by
permission of Kluwer Academic Publishers).

indeed the case then the critical conditions for ductile crack growth, in thin-sheet
materials of relatively high ductility, can be given in terms of the strain ϵ_y and strain
rate $\dot{\epsilon}_x$, on the axis of symmetry ahead of the crack tip, by the localised-necking condition (7.13) and the parabolic-characteristic condition ($\dot{\epsilon}_x = 0$) written in the following form:-

$$\epsilon_y^L \geqslant m ,\tag{8.12a}$$

$$\dot{\epsilon}_x = 0 ;\tag{8.12b}$$

where m is the work-hardening exponent. When these conditions are satisfied
intense localised necking can develop ahead of the crack tip; it will be recognised
that both conditions can be readily monitored in experiments.

8.5 Ductile Fracture at the Root of Cracks under Large-Scale Yielding Conditions.

With materials of high fracture toughness the critical conditions for the catastrophic propagation of a crack will usually occur only after the development of large-scale plasticity in the crack-tip region. Under these conditions localised ductile fracture by microvoid nucleation, growth and coalescence develops in a stable manner at the crack tip to give extensive sub-critical crack growth. These effects greatly complicate the ductile fracture problem since it now becomes necessary to characterise the onset of large scale instability in the ductile crack-growth process. In recent years attempts have been made to characterise the ductile crack-growth process in terms of the J-integral parameter [22], but it is becoming increasingly clear that the concept of J-controlled crack growth gives an inadequate description of the critical conditions leading to a transformation from stable sub-critical crack growth to catastrophic fracture.

The fundamental problem with J-integral theory lies in its formulation in terms of a 'deformation' theory of plasticity [22] rather than the more acceptable 'incremental' or 'flow' theory of plasticity [1,4] described in Chapter 1.2. A deformation theory of plasticity can form an adequate model of a real elastic/plastic work-hardening metal only under a highly restricted set of conditions consisting of proportional loading under monotonically increasing loads and no unloading [1,4]. To illustrate the restricted validity of the deformation theory we begin with the Prandtl-Reuss equations (1.22), written in terms of the principal components of deviatoric stress S_i and plastic strain-increment $d\epsilon_i^p$ in the form:-

$$d\epsilon_i^p = d\lambda S_i ; \qquad (8.13)$$

where $d\lambda$ is a scalar factor of proportionality. Now proportional loading is defined by the following relation between the current deviatoric stress S_i and its initial value S_i^o:-

$$S_i = P S_i^o ; \qquad (8.14)$$

where P is a monotonically increasing parameter. It follows from equation (8.14) that all states of proportional loading are represented by straight radial lines on the π - plane projection of the yield surface (cf. Chapter 1.2 and Fig. 1.3(b)), running from the origin through the initial yield stress state S_i^o, and the subsequent loading surfaces, Fig. 8.14(a). If we assume that the principal axes remain fixed in direction, the proportional-loading equation (8.14) can be substituted into equation (8.13) and the resulting expression integrated [1,4] to give the following relation between the total plastic strain ϵ_i^p and the current deviatoric stress S_i :-

$$\epsilon_i^p = \left(\frac{\int Pd\lambda}{P} \right) S_i . \qquad (8.15)$$

This is the Hencky deformation theory of plasticity which immediately becomes invalid [1,4] if the stress and strain paths show any deviation from the purely radial

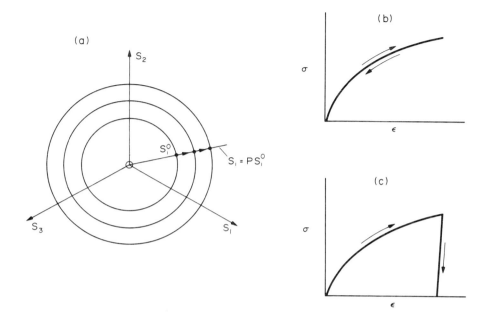

Fig. 8.14. (a) A proportional-loading path on the π-plane projections of
the yield and loading surfaces of a work-hardening material.
The effects of loading and unloading for materials obeying :- (b)
a deformation theory of plasticity; (c) a flow theory of plasticity.

path (Fig. 8.14(a)) of proportional loading or if any unloading occurs (i.e. dP < 0).
Equation (8.15) is strictly equivalent to a theory of non-linear elasticity rather than
a theory of plasticity and the wide disparity which develops between the 'deformation' and 'flow' theories of plasticity, when unloading occurs, is illustrated
schematically in Figs. 8.14(b) and (c). It is clear therefore that, if the J-integral is to
represent an acceptable parameter for describing the sub-critical crack growth process, the conditions of proportional loading and no-unloading of elements in the
crack-tip plastic field must at the very least be satisfied predominantly throughout
the plastic field. Unfortunately, despite the conditions for the validity of J-controlled crack growth proposed by Hutchinson and Paris [23], it can be shown [24] that
the J-integral is strictly invalid under conditions of both small-finite crack growth
and infinitesimal crack growth. Since a loss of validity under infinitesimal crack
growth is sufficient to invalidate the J-integral under any degree of sub-critical
crack growth we consider below only the case of infinitesimal growth.

Attempts have been made to justify the use of a deformation theory of plasticity on the grounds that, if a yield-vertex effect were to develop, then departures from proportional loading would show little difference between the predictions of deformation or flow theories of plasticity. However, as pointed out in Chapter 1.3, there is no experimental evidence to support the existence of deviatoric yield vertices and results from the physical theory of plasticity suggest that if yield vertices were to exist at all they would be eliminated by a plastic strain-increment as small as 1% of the elastic yield point strain.

8.6   The Validity of the J-Integral under Conditions of Infinitesimal Crack Growth.

The procedure of Hutchinson and Paris [23] for examining the validity of the J-integral under conditions of small stable growth was based on the power-law hardening solutions for crack-tip fields, derived from a small-strain deformation plasticity theory by Hutchinson [25,26] and Rice and Rosengren [27]. For a deformation theory of plasticity the total plastic strain ϵ_{ij}^P is given in terms of the stress σ_{ij} by :-

$$\epsilon_{ij}^P = \alpha \, 3/2 \, (\bar{\sigma})^{m-1} \, S_{ij} \tag{8.16}$$

where $S_{ij} = \sigma_{ij} - \frac{1}{3} \delta_{ij} \sigma_{kk}$ is the deviatoric stress, $\bar{\sigma}$ is the equivalent stress, m is the power-hardening exponent and α is a dimensionless constant. The corresponding dominant singularity field at a crack-tip, in terms of a polar coordinate system (r,θ), is then given for a far-tension perpendicular to the crack in the form:-

$$\sigma_{ij}\,(r,\theta) = k_n(J/r)^{1/(m+1)} \, \tilde{\sigma}_{ij}\,(\theta) ,$$

$$\epsilon_{ij}^P\,(r,\theta) = \alpha k_n(J/r)^{m/(m+1)} \, \tilde{\epsilon}_{ij}^P\,(\theta) , \tag{8.17}$$

$$\bar{\sigma}\,(r,\theta) = k_n(J/r)^{1/(m+1)} \tilde{\bar{\sigma}}\,(\theta) ,$$

where k_n is a constant and the functions $\tilde{\sigma}_{ij}(\theta)$, $\tilde{\epsilon}_{ij}^P(\theta)$, $\tilde{\bar{\sigma}}(\theta)$ are given for both plane strain and plane stress by Hutchinson [25,26].In the immediate vicinity of the crack tip the plastic strains greatly exceed the elastic strains and it can be shown that, for a stationary crack under monotonically increasing loads, the stresses are proportional [28]; the results in equation (8.17) will then be identical to those obtained from the incremental (flow) theory of plasticity.

In an attempt to justify the validity of equations (8.17) under conditions of sub-critical crack growth, and thus introduce the concept of J-controlled crack growth, Hutchinson and Paris [23] obtain the first differential $d\epsilon_{ij}^P$ and use the expression to suggest that proportional loading will occur if $dJ/J \gg da/r$; they also suggest that elastic unloading will be confined to a radius from the crack tip of the order of the crack extension Δa. Unfortunately, in deriving these conditions for the validity of

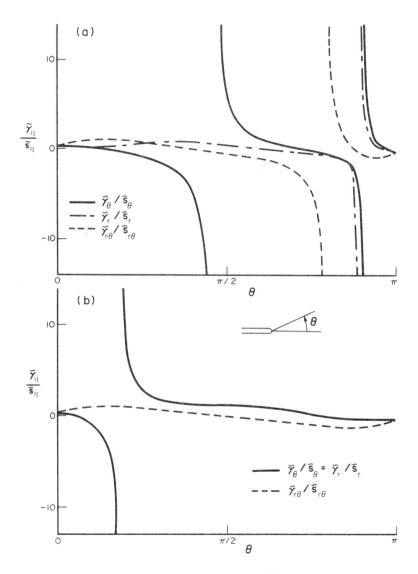

Fig. 8.15. The non-proportional loading term $\tilde{\gamma}_{ij}/\hat{S}_{ij}$ in the deviatoric-stress differential for a power-hardening exponent m = 3; (a) plane-stress and (b) plane-strain. (After P.F. Thomason, Int. J. Fract., 1989, in press. Reprinted by permission of Kluwer Academic Publishers).

J-controlled crack growth, Hutchinson and Paris neglect the presence of zeros in the deviatoric-stress functions $\tilde{S}_{ij}(\theta)$, which cause a vast deviation from proportional loading in wide fan-shaped regions spreading across the stress field of the dominant singularity [24]; the concept of J-controlled crack growth is therefore rendered invalid even for infinitesimal crack growth. To show these non-proportional loading effects we consider directly the differential of the deviatoric stress which can be written in the form:-

$$\frac{dS_{ij}}{S_{ij}} = \frac{1}{m+1} \frac{dJ}{J} + \frac{da}{r} \frac{\tilde{\gamma}_{ij}(\theta)}{\tilde{S}_{ij}(\theta)} \, , \qquad (8.18)$$

where

$$\tilde{\gamma}_{ij}(\theta) = \frac{1}{m+1} \tilde{S}_{ij}(\theta) \cos\theta + \sin\theta \frac{\partial \tilde{S}_{ij}(\theta)}{\partial \theta} \, . \qquad (8.19)$$

The functions $\tilde{S}_{ij}(\theta)$ are given by Hutchinson [25,26] for plane stress and plane strain and a typical set of results [24] for $\tilde{\gamma}_{ij}(\theta)/\tilde{S}_{ij}(\theta)$ is given in Fig. 8.15 and 8.16 for m = 3 and m = 13, respectively. These results show that there is a very large departure from proportional loading in a wide region around the zeros of $\tilde{S}_{ij}(\theta)$ and the locations of the zeros are not confined to the crack-wake ($\pi/2 \leqslant \theta \leqslant \pi$) regions; the results also show that the non-proportional loading effect is intensified with increase in the power-hardening exponent m. The non-proportional loading results show that there will be wide fan-shaped regions spreading completely across the stress field of the dominant singularity where the deformation theory of plasticity, on which the J-integral theory is based, is invalid. This result is also confirmed in an analysis of finite crack growth [24] which also shows that elastic-unloading effects extend across the stress field of the dominant singularity up to distances more than fifty times greater than the finite crack-extension Δa. It must be concluded therefore that the concept of J-controlled crack growth is invalid.

There is therefore an urgent need to develop a new theory for dealing with problems involving extensive ductile crack growth prior to catastrophic fracture, as observed in materials of high fracture toughness. The new crack-growth theory must clearly be based on an incremental or 'flow' theory of plasticity which is not invalidated by either non-proportional loading or elastic-unloading effects; a recent theory of this type [24] is briefly described in the following section.

8.7 The Crack-Stability Problem for an Incremental-Plastic/Elastic Solid

In formulating the crack stability problem for a work-hardening plastic/elastic solid, obeying the incremental or flow theory of plasticity, we consider the case [24] where a crack has already undergone extensive sub-critical growth following the application of prescribed tractions F_i over the surface S_F and prescribed displacements u_i over

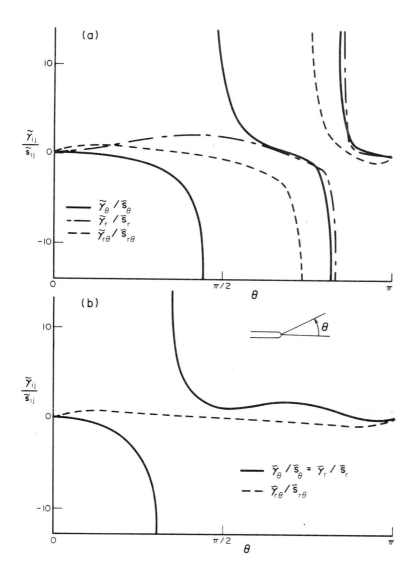

Fig. 8.16. The non-proportional loading term $\tilde{\gamma}_{ij}/\check{S}_{ij}$ in the deviatoric-stress differential for a power-hardening exponent m = 13; (a) plane-stress and (b) plane-strain. (After P.F. Thomason. Int. J. Fract. 1989, in press. Reprinted by permission of Kluwer Academic Publishers).

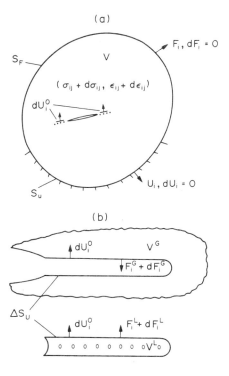

Fig. 8.17. (a) The general incremental-plastic/elastic body containing a
crack and subject to dead loading on S_F, rigid constraints on
S_u and a virtual displacement-increment field satisfying the
boundary conditions and giving du_i^o on ΔS_u.
(b) The imaginary 'cutting' operation ahead of the crack tip
separating the cracked body into local and global volumes
with a common interface ΔS_u. (After P.F. Thomason, Int. J.
Fract., 1989, in press. Reprinted by permission of Kluwer
Academic Publishers).

the remaining surface S_u; Fig. 8.17. We now set out to establish a sufficient condition for the *stability* of the current equilibrium state in the cracked body V, with
dead loading ($dF_i = 0$ on S_F) and rigid constraints ($du_i = 0$ on S_u), using an incre-
mental-displacement formulation in which the linear equations of solid mechanics
remain valid [24].

The current equilibrium state can be regarded as definitely *stable* if it satisfies the
Kelvin-Dirichlet criterion of stability [29] for the set of virtual displacement-increment fields corresponding to the formation of *incipient* fracture damage in a fracture-

process zone ahead of the crack tip (Fig. 8.17). The sufficient condition for stability has the form $\delta(I-E) > 0$ for all virtual displacement-increment fields satisfying the boundary conditions; where δI is the internal energy stored or dissipated and δE is the work done by the external dead loading. For a work-hardening plastic/elastic solid the expressions for δI and δE are equal to first order and it is necessary to compare second-order expressions in the stability condition. Appropriate second-order expressions for δI and δE can be obtained by an imaginary cutting and rewelding operation in the vicinity of the crack tip [24] to separate the total volume V into a global volume V^G, where the incremental response will be primarily influenced by the geometry of the body, and a local volume V^L where the incremental response will be primarily influenced by the plastic flow and fracture characteristics of the material, Fig. 8.17. The 'separate' volumes V^L and V^G are now subjected to prescribed virtual displacement-increment fields, giving the displacement-increments du_i^o on the common surfaces ΔS_u of the two volumes (Fig. 8.17); thus the common surfaces remain in perfect contact when the virtual displacement-increment field is applied. Now the equilibrium stress field in V^G will change from σ_{ij}^G to $(\sigma_{ij}^G + d\sigma_{ij}^G)$ and the tractions on ΔS_u from F_i^G to $(F_i^G + dF_i^G)$; on ΔS_u of V^L the tractions will change from F_i^L to $(F_i^L + dF_i^L)$. Hence, we can write the Kelvin-Dirichlet stability criterion $(\delta I - \delta E > 0)$ for the complete cracked body $(V^G + V^L)$ in the following form [24]:-

$$\int_{V^G} (\sigma_{ij}^G + \tfrac{1}{2}\, d\sigma_{ij}^G)\, d\epsilon_{ij}^G\, dV + \int_{V^L} (\sigma_{ij}^L + \tfrac{1}{2}\, d\sigma_{ij}^L)\, d\epsilon_{ij}^L\, dV - \int_{S_F} F_i^G\, du_i^G\, dS > 0 . \qquad (8.20)$$

Using the virtual-work relations for the *separate* equilibrium stress fields in V^G and V^L the volume integrals in (8.20) can be transformed to give the following sufficient condition for stability of the current equilibrium state of a cracked body [24]:-

$$\tfrac{1}{2} \int_{\Delta S_u} (dF_i^L + dF_i^G)\, du_i^o\, dS > 0 ; \qquad (8.21)$$

where $F_i^L + F_i^G = 0$ on ΔS_u. It is important to note here that at the stability limit, where condition (8.21) *just* fails to be satisfied, there is no implication that $dF_i^L + dF_i^G = 0$ at all points on ΔS_u; ie the stability limit is not identified as an adjacent position of equilibrium.

If the sufficient condition for stability (8.21) is satisfied for the complete set of virtual displacement-increment fields du_i, giving du_i^o on ΔS_u, the crack cannot grow under dead loading and rigid constraints since none of the virtual fields could become actual fields in a body that has been 'rewelded' on ΔS_u. The current crack length is therefore *stable* and crack growth can only occur under these conditions if either a traction-increment is applied to S_F or a displacement-increment to S_u. If the stability condition (8.21) continues to be satisfied after each such increment of crack growth the resulting situation is mathematically equivalent to stable crack growth.

Ductile Fracture of Metals

In order to develop a method for applying the crack-stability criterion (8.21) in practical problems, the inequality can be rewritten in the form:-

$$M + S > 0 ,$$ (8.22)

where

$$M = \frac{1}{2} \int_{\Delta S_u} dF_i^L \, du_i^o \, dS$$ (8.23)

and

$$S = \frac{1}{2} \int_{\Delta S_u} dF_i^G \, du_i^o \, dS .$$ (8.24)

The M integral represents a component of the total energy *increment* applied to ΔS_u of V^L and is related to the *gradient* of an effective R - curve for the material; it can therefore be regarded as an autonomous material response [24]. On the other hand the S integral represents a component of the total energy increment applied to ΔS_u of V^G and this will be primarily dependent on the particular geometry of the cracked body. It should be noted that the M and S integrals are not path independent and the size and shape of V^L must therefore be fixed for a given problem [24].

The proposed incremental criterion of crack-stability (8.21) requires extensive experimental development, and a preliminary experimental procedure for measuring the M integral and evaluating the S integral in practical problems has been described elsewhere [24].

REFERENCES

1. Hill, R., *The Mathematical Theory of Plasticity,* Clarendon Press, Oxford, 1950.
2. Johnson, W., Sowerby, R., and Haddow, J.B., *Plane-Strain Slip-Line Fields,* Edward Arnold, London, 1970.
3. Griffis, C.A., and Spretnak, J.W., *Trans. Iron and Steel Inst. Japan,* 1969, 9, p.372.
4. Kachanov, L.M., *Fundamentals of the Theory of Plasticity,* MIR Publishers, Moscow, 1974.
5. Green, A.P., *Quart. J. Mech. and Appl. Maths.,* 1953, 6, p.223.
6. Alexander, J.M., and Komoly, T.J., *J. Mech. Phys. Solids,* 1962, 10, p.265.
7. Green, A.P., *J. Mech. Phys. Solids,* 1956, 6, p.259.
8. Ewing, D.J.F., *J. Mech. Phys. Solids,* 1968, 16, p.205.
9. Knott, J.F., *Fundamentals of Fracture Mechanics,* Butterworths, London, 1973.
10. Hellan, K., *Introduction to Fracture Mechanics,* McGraw-Hill, New York, 1985.
11. Ewalds, H.L., and Wanhill, R.J.H., *Fracture Mechanics,* Edward Arnold, London, 1984.
12. Thomason, P.F., *Int. J. Fract. Mech.,* 1971, 7, p.409.
13. Goods, S.H., and Brown L.M., *Acta Metall,* 1979, 27, p.1.
14. Le Roy, G., Embury, J.D., Edwards, G., and Ashby, M.F., *Acta Metall,* 1981, 29, p.1509.

15. Wilshaw, T.R., Rau, C.A., and Tetelman, A.S., *Eng. Fract. Mech.,* 1968, 1, p.191.
16. Creager, M., and Paris, P.C., *Int. J. Fract. Mech.,* 1967, 3, p.247.
17. Thomason, P.F., *Fracture Mechanics in Engineering Application,* (edited by G.C. Sih and S.R. Valluri), p.43, Sijthoff and Noordhoff, The Netherlands, 1979.
18. Nishimura, N. and Achenbach, J.D., *J. Mech. Phys. Solids.,* 1986, 34, p.147.
19. Hutchinson, J.W., *J. Mech. Phys. Solids,* 1968, 16, p.337.
20. Hill, R., *J. Mech. Phys. Solids,* 1952, 1, p.19.
21. Hahn, G.T., Mukherjee, A.K., and Rosenfield, A.R., *Eng. Fract. Mech.,* 1971, 2, p.273.
22. Rice, J.R., *J. Appl. Mech.,.* 1968, 35, p.379.
23. Hutchinson, J.W., and Paris, P.C., *Elastic-Plastic Fracture,* A.S.T.M., S.T.P., 668, p.37, American Society for Testing Materials, 1979.
24. Thomason, P.F., *Int. J. Fract.* 1989, in press.
25. Hutchinson, J.W., *J. Mech. Phys. Solids,* 1968, 16, p.13.
26. Hutchinson, J.W., *J. Mech. Phys. Solids,* 1968, 16, p.337.
27. Rice, J.R., and Rosengren, G.F., *J. Mech. Phys. Solids,* 1968, 16, p.1.
28. Goldman, N.L., and Hutchinson, J.W., *Int. J. Solids and Struct.,* 1975, 11, p.575.
29. Hill, R., *J. Mech. Phys. Solids,* 1959, 7, p.209.

INDEX